PHYSICAL DESIGN FOR
MULTICHIP MODULES

THE KLUWER INTERNATIONAL SERIES
IN ENGINEERING AND COMPUTER SCIENCE

PHYSICAL DESIGN FOR MULTICHIP MODULES

by

M. Sriram
Beckman Institute for Advanced Science & Technology
University of Illinois, Urbana-Champaign

S. M. Kang
Coordinated Science Laboratory
University of Illinois, Urbana-Champaign

KLUWER ACADEMIC PUBLISHERS
Boston / Dordrecht / London

Distributors for North America:
Kluwer Academic Publishers
101 Philip Drive
Assinippi Park
Norwell, Massachusetts 02061 USA

Distributors for all other countries:
Kluwer Academic Publishers Group
Distribution Centre
Post Office Box 322
3300 AH Dordrecht, THE NETHERLANDS

Library of Congress Cataloging-in-Publication Data

A C.I.P. Catalogue record for this book is available
from the Library of Congress.

Printed on acid-free paper.

Printed in the United States of America

To my parents, Jayaram and Shyamala, and my wife Deepa

- M. Sriram

To my wife Mia and my children, Jennifer and Jeffrey

- S. M. Kang

CONTENTS

PREFACE

The density and performance of VLSI chips have now increased to a point where further improvements in system speed and size are limited by packaging. This has become increasingly evident in recent years, and considerable effort is now being expended to develop new packaging technologies.

Multichip modules (MCMs) have received a lot of attention lately, as a possible solution to the packaging bottleneck. They have been demonstrated to be able to meet the requirements of high density, high performance and low cost, while providing a number of additional benefits as well. Most chip and system manufacturers are either actively involved in MCM technology, or are considering its usage.

While MCMs provide a number of significant advantages over conventional packaging, they also introduce some difficult design, manufacturing and testing problems. The development of new materials with superior electrical, mechanical and thermal properties is one of the biggest challenges. Effective multichip system design also requires the consideration of packaging effects, such as heat dissipation and transmission line effects, during all stages of the system design process.

Physical design is a very significant part of the system and package design process. Broadly speaking, physical design is the process of laying out the components of a system and establishing electrical connections between them, subject to various physical, electrical and thermal constraints.

VLSI physical design algorithms have been a very important field of research in the past 20–30 years. These algorithms have contributed significantly to the rapid improvements in VLSI chips. Unfortunately, the tools and algorithms that have been developed for VLSI physical design cannot be directly used for designing MCMs. The physical constraints in MCM substrates differ considerably from VLSI, and the electrical and thermal requirements are also more stringent in MCMs.

VLSI physical design problems, at least until recently, have been viewed as combinatorial or geometrical problems, and have been solved using techniques such as linear programming and graph algorithms, with little or no regard for the complex electrical behavior of the objects being manipulated. These approaches make approximations which are reasonable for VLSI design, but are unacceptable for MCM design. For example, in VLSI design, the delay of an interconnection net is commonly assumed to be proportional to the total capacitance (and hence total length) of the net. Based on this assumption, a number of delay minimization algorithms have been developed. However, none of these can be directly applied to MCMs, where the resistance, inductance, capacitance and topology of an interconnection all affect its delay significantly.

System designers are realizing that, in order to maximize the benefit of using MCMs, they need CAD tools developed specifically for MCMs. A number of CAD vendors and researchers at universities are now involved in developing MCM-specific approaches and design tools. This book is intended to provide these professionals with information on the physical and electrical constraints unique to MCM design, as well as to provide them with a comprehensive survey of existing state-of-the-art MCM physical design algorithms.

This book can also be used as a textbook for an advanced graduate level course in physical design, or as a reference volume for graduate students conducting research in the areas of MCM physical design and interconnect analysis. A large amount of useful information on MCM technology and manufacturing processes has been collected from various sources, and presented with an emphasis on the impact on physical design. The readers of this book can thus avoid having to read a number of articles with different emphases in order to gain an insight into the physical nature of MCMs. A comprehensive bibliography is provided for those requiring more detailed information on specific topics. The material is also interspersed with a number of open problems and suggestions for future research, which will be helpful to researchers and graduate students looking for timely and interesting research topics.

Overview

The first chapter of this book is intended as an introduction to multichip modules. It describes the MCM technology alternatives available today, and discusses the relative merits of each approach. The advantages of MCMs and their impacts on system performance are described, and the challenges faced by MCM system designers are discussed.

The electrical performance of an MCM is defined by the behavior of its interconnects. Due to the particular dimensions and operating frequencies of MCMs, the interconnects need to be treated as lossy coupled transmission lines. Chapter 2 surveys a number of approaches developed recently for rapid approximation of the time domain response of such interconnects. These approaches yield good approximations of interconnect response with orders of magnitude speedup over conventional simulation programs, and are thus very useful during physical design.

Delay models for MCM interconnects are also presented in Chapter 2. RC delay models suitable for lower-speed applications, as well as a new second-order RLC delay model for high-speed interconnects, are presented.

Chapter 3 discusses the system partitioning and chip placement problems. Performance-driven partitioning algorithms are described. The differences between VLSI and MCM placement are clearly outlined, and new algorithms for placement are presented, which take into account the effect of interconnect resistance on delay. Recent work on layout problems for opto-electronic MCMs is also described.

Chapter 4 deals with the multilayer routing problem in MCM substrates. There are many possible approaches to this problem, and these are discussed in detail. Some of the popular recent algorithms for MCM multilayer routing are described.

Chapter 5 discusses a different approach to the multilayer routing problem, in which the three-dimensional routing problem is decomposed into two stages: two-dimensional tree construction and layer assignment. Recent work on performance-driven tree construction algorithms is surveyed, and algorithms for minimizing transmission-line effects and second-order delays are described. Specialized routing algorithms for clock trees are also discussed. The layer assignment problem in different MCM environments is investigated in Chapter 6, and different formulations and algorithms for the problem are described in detail.

The last chapter summarizes the material in the book, and collects together the various open problems and suggestions for future work found in the previous chapters.

ACKNOWLEDGMENTS

The material presented in this book consists of original research work, as well as an extensive survey of published material in the areas of physical design and multichip modules. We would like to take this opportunity to acknowledge the individuals and organizations who contributed to this book.

The original research presented in this book was supported by a number of research grants. The authors would like to express their gratitude to Dr. C. K. Wong at the IBM T. J. Watson Research Center, Yorktown Heights, and to IBM Corporation, for supporting the first author through an IBM Computer Sciences Fellowship from 1990 – 1993. The research was also funded in part by the Joint Services Electronics Program, the Semiconductor Research Corporation, the National Science Foundation and the Illinois Technology Challenge Grant.

Professors C. L. Liu, Timothy Trick and David Wilcox at the University of Illinois at Urbana-Champaign made many useful suggestions during the course of this work. The authors also acknowledge the contributions of Professor Majid Sarrafzadeh and Dr. Jun-Dong Cho at Northwestern University, for their collaboration and many fruitful discussions. Many people have made outstanding contributions to the field of MCM physical design, but the constraints of time and space have prevented us from including all these results in this book. Nevertheless, we extend our appreciation to all of these dedicated researchers.

Finally, we would like to express our deep gratitude to our respective wives, Deepa and Mia, whose support and encouragement made it possible for us to successfully complete this project.

M. Sriram
S. M. Kang

Urbana, Illinois

PHYSICAL DESIGN FOR MULTICHIP MODULES

1

INTRODUCTION

1.1 THE PACKAGING BOTTLENECK

Ever since its invention, the density and performance of the integrated circuit have been increasing at an exponential rate, fueled by improvements in processing, devices, circuit design and layout techniques. This has led to dramatic improvements in the size, speed and cost of electronic systems. Unfortunately, not all of the components of these systems have kept pace with the integrated circuit. The speed and density of today's VLSI chips are so high that IC packages now present a bottleneck for increasing system speed and shrinking system size.

Conventional packaging technology consists of several levels of packaging hierarchy (Fig. 1.1). On the first level, bare dies are packaged into plastic or ceramic single-chip packages (SCPs). These packages are then mounted onto printed-circuit boards (PCBs) using either through-hole or surface-mount technologies. The PCBs may in turn be connected together using higher-level packages, such as backplanes. In general, each successive level of packaging has larger feature sizes, and thus results in unwanted reduction in packing density, performance and reliability, and increase in system size, weight and power consumption.

The conventional SCP-on-PCB approach results in enormous inefficiency in size and performance. On a typical printed circuit board with several VLSI chips, the *silicon efficiency* (the ratio of the total area occupied by silicon to the total board area) is typically around 5%. Figure 1.2(a) shows a portion of a PCB with 10 SCPs mounted on it. The approximate sizes of the silicon chips inside the single-chip packages are shown by dark rectangles. Figure 1.2(b) shows how much the area of the board can be reduced if the single chip packages

1

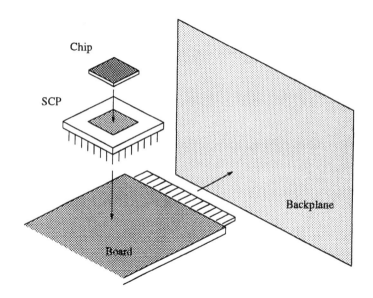

Figure 1.1   Conventional packaging hierarchy

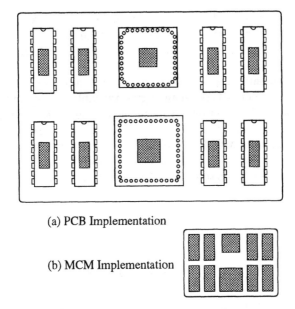

(a) PCB Implementation

(b) MCM Implementation

Figure 1.2 Area savings by eliminating single chip packages

are eliminated, and the chips are mounted directly onto the board [1]. The area savings can be an order of magnitude or more. This idea is the essence of the *multichip module* (MCM): to eliminate the single-chip package, and attach several bare chips directly onto one package.

Designers of high-performance systems have long known that packaging plays a critical role in determining system speed. For example, IBM has been using multilayer ceramic "Thermal Conduction Modules" (TCMs), housing approximately 100 bare dies, for well over a decade in its mainframe computers. However, until recently, these expensive packages were limited exclusively to the domain of high-end systems. Conventional packaging provided adequate performance and density at much lower cost, for all but the most demanding applications.

In recent years, that situation has changed considerably, for a number of reasons. The most important reason is the fact that conventional packaging has not been able to keep pace with rapid advances in the performance and density of integrated circuits. Packaging is now the limiting factor impeding further advances in system speed and density.

Portable computing is another force driving advances in packaging. Size, weight, power consumption and ruggedness are all critical factors for portability, and none of these purposes is served satisfactorily by conventional packages.

This chapter presents an overview of multichip packaging. The next section describes the different MCM technologies available today, and their relative merits and drawbacks. Section 1.3 discusses the benefits of MCM technology, which extend far beyond the obvious reduction in size. Section 1.4 describes some of the challenges MCM and system designers have to overcome in order to generate reliable, working designs, and reviews some of the research projects and commercially available tools for MCM design. Section 1.5 gives an overview of this book.

1.2 MCM TECHNOLOGIES

At present, MCMs are classified into three major categories, which differ mainly in the materials used to fabricate the interconnection layers and substrate. In addition, there are at least four major technologies available for bonding chips to the package. This section describes the salient features of these technology options.

1.2.1 MCM substrate/interconnect technologies

Based on the type of material used in the interconnection wiring layers, MCMs are classified as *MCM-C* (multilayer ceramic), *MCM-L* (laminated) or *MCM-D* (deposited thin-film).

MCM-C

Ceramic multichip module (MCM-C) technology has been in use for high-performance computer packaging for over a decade [2]. A cross-section of a MCM-C substrate is shown in Fig. 1.3. Chips are bonded to the top layer of a multilayer ceramic substrate. Each layer is patterned with metal wiring for routing interchip connections, connections to module I/O pads, and for supplying power and ground to the chips. Module I/O pads may be brazed to

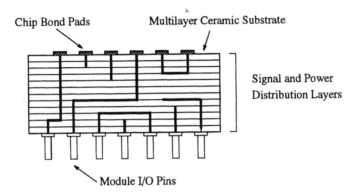

Figure 1.3   Multilayer ceramic MCM

the bottom layer, or may be arranged around the periphery of the module as surface-mount pads.

Ceramic modules are constructed using layers of "greensheet," which is a mixture of ceramic and glass powder suspended in an organic binder [3]. These layers are customized by punching holes for the vias and forming wiring patterns using molybdenum paste. The layers are then stacked together and baked to obtain a multilayer alumina substrate. Recent improvements in the technology include glass-ceramic substrates and copper metallization. The use of glass-ceramic substrates allows increased operating speeds, due to the lower dielectric constant (5.0 as compared to 9.4 for alumina). It also increases module reliability by providing a better thermal expansion coefficient match to silicon. Chips are bonded to the top layer of the substrate, typically using flip-chip bonding. Due to the relatively low wiring density achievable with the screening and punching processes used to form the interconnect wiring, many wiring layers may be required to complete the routing of a complex design: as many as 63 layers are used in IBM's recent ceramic module [4]. The advantages of MCM-C include the excellent dimensional stability and mechanical properties of ceramics, which lead to high reliability and module power dissipation capacity. The maturity of this technology also allows it to compete favorably in cost with other MCM technologies.

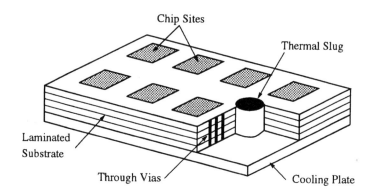

Figure 1.4 Laminated MCM substrate

MCM-L

Laminated MCMs are essentially extensions of PCB technology to finer line widths. A typical MCM-L substrate is illustrated in Fig. 1.4. Currently, MCM-L substrates can provide line widths and spacings as fine as 75 microns (3 mils), which is sufficient for a number of low-cost applications. At present, MCM-L is the most cost-effective technology for applications which do not require very large routing resources, or demand very high performance, since it is based on PCB technology which has matured over several decades. Substrate costs increase significantly as line widths become finer. There are on-going efforts to push MCM-L technology to 2 mil and finer line widths and spacings and to provide smaller via dimensions and blind and buried vias at low cost. Thermal performance is poorer in MCM-L, due to the lower thermal conductivity of the substrate. Thermal vias and slugs can be used to improve heat removal [5], as illustrated in the figure.

MCM-D

Deposited thin-film MCMs consist of alternate thin-film dielectric and metallization layers deposited on a substrate, which may be ceramic, silicon or aluminum (Fig. 1.5). The dielectric materials include polyimide and silicon dioxide, while the metallization is usually aluminum or copper. The interconnect technology is very similar to on-chip interconnect, providing as much as 1000 inches/inch^2 of wiring capacity. At present, this technology is not very cost-effective, due to the lack of a vendor infrastructure and the high processing

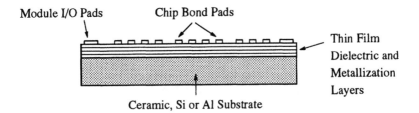

Module I/O Pads   Chip Bond Pads

Thin Film
Dielectric and
Metallization
Layers

Ceramic, Si or Al Substrate

Figure 1.5   Deposited thin-film MCM substrate

cost. The wiring capacity greatly exceeds the requirements of most systems - MCM-D designs rarely use more than two signal wiring layers. However, as operating speeds go beyond 200 MHz, MCM-D may become necessary.

1.2.2   Chip bonding technologies

The technique used to bond bare dies to the MCM affects the maximum achievable silicon efficiency, as well as electrical, mechanical and thermal properties of the package. Four popular bonding techniques are described below; the list will surely grow as more efficient and reliable techniques are invented.

Wirebonding

Wirebonding is the traditional bonding technique used in SCPs. Chips are attached to the substrate, using a special thermally conductive adhesive, with the I/O pads facing up. Fine wires, usually made of gold, are attached to the chip I/O pads and corresponding metallized pads on the substrate, using ultrasonic or thermocompression bonding (Fig. 1.6). The primary advantage of this technique is that it does not require any new equipment for MCMs: the SCP wirebonding equipment can be used. Thus, it is cost-effective for applications which do not require very high density or performance. Since the connections are formed one at a time, this technique is inefficient for chips with very high pin counts. I/O pads must be located on the perimeter of the chip: area arrays cannot be handled. The parasitics of the bonding wires may introduce signal degradation in high-speed designs.

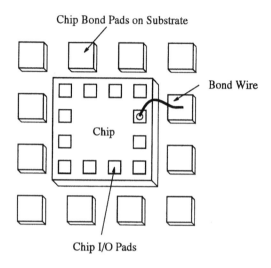

Figure 1.6 Wirebonding

Tape Automated Bonding

In Tape Automated Bonding (TAB), bare dies are first bonded to a carrier, which is a long roll of flexible tape similar to photographic film (Fig. 1.7). The tape has metallized contact pads and wiring patterns deposited on it. Once the dies are bonded to the tape by "inner-lead bonding," the individual chip carriers can be cut out of the tape, and pads on the TAB carrier can be bonded to pads on the substrate ("outer lead bonding"). The TAB process is easy to automate, and lends itself to volume production. One of the most significant advantages is that after inner lead bonding, the bare dies can be tested and burned-in, since the carrier provides easy access to the chip pads. This greatly increases the confidence level in the bare dies and the module yield. However, the silicon efficiency is not as high as other techniques, due to the space occupied by the TAB leads.

Flip-chip

"Flip-chip" technology, which is sometimes known as *Controlled Collapse Chip Connection* (C4), attaches bare dies to metallized substrates by forming solder bumps on the chip I/O pads, placing the chips face down on the substrate so that the bumps align with contact pads on the substrate surface, and reflowing

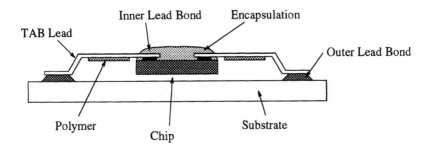

Figure 1.7 Tape automated bonding

the solder to simultaneously form all the contacts (Fig. 1.8). This technology has been used by IBM in its Thermal Conduction Modules (TCMs). This approach has some significant advantages – the silicon efficiency is very high, and unlike wirebond and TAB, there are no leads between the chip and package, so electrical parasitics are very low. The process exhibits a "self-alignment" characteristic due to the surface tension of solder: small misalignments in the placement of the chip are automatically corrected. Unlike wirebonding, where the connections are processed serially, all the connections are completed in one step. Furthermore, the process can handle chips with area arrays of I/O pads. The last two factors are essential for handling chips with very high pin counts. Removal of faulty chips is possible without damaging the chip site bond pads. On the down side, heat removal becomes difficult, since the back side of the chip is not in contact with the substrate. Complex heat removal mechanisms, such as spring-loaded copper pistons contacting the back sides of the chips, have been developed for high-power modules [4].

Overlay

A relatively recent approach to MCM interconnect is the "overlay" interconnect technology [6]. Instead of placing the chips on top of a prefabricated interconnection structure, the chips are placed first in milled cavities in a ceramic substrate, and the interconnect structure, composed of thin-film dielectric and metallization layers, is fabricated over the chips (Fig. 1.9). The process is analogous to the IC fabrication process. Silicon efficiency of as much as 90% has been reported using overlay interconnect, since chips can be placed with very little separation [7]. Rework of faulty modules can be accomplished by removing the interconnect structure, replacing defective parts, and repeating the fabrication process. High density, reliability and performance are some of the important features of this process.

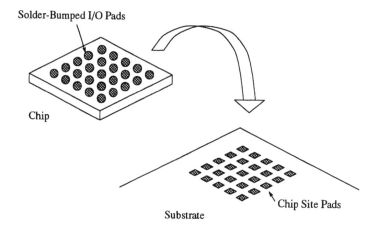

Figure 1.8 Flip-chip bonding

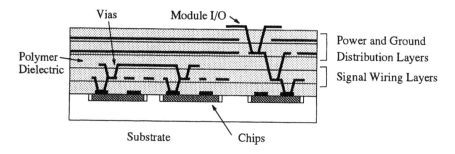

Figure 1.9 Overlay or chips-first interconnect

1.3   ADVANTAGES OF MULTICHIP PACKAGING

Some of the advantages of multichip technology are illustrated in Fig. 1.10. The graph also shows how some of these benefits are related – for example, increased performance is due to reduced dimensions as well as reduced parasitics. Each of these effects and relationships is described in detail below.

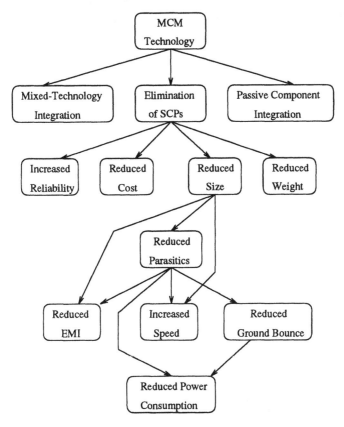

Figure 1.10   Advantages of multichip packaging

1.3.1   System size and weight

The impact of MCMs on the physical dimensions and weight of a system is best illustrated by the digital signal processor (DSP) module developed at

Rockwell International for the Eglin Air Force Base under the Signal Processor Packaging Design (SPPD) program [8]. The design of the signal processor was done in two phases: in the first phase, the focus was on hardware and software design issues, and not on packaging. The design was implemented on a stack of four circuit boards. In the second phase, the focus was entirely on packaging. The design was implemented on a multichip module with a thin-film interconnect substrate. The savings in package weight and volume are enormous: the PCB implementation measured 26.7 cm × 19.1 cm × 8.3 cm, occupying a volume of 4233 cm^3. The MCM, on the other hand, measured only 8.3 cm × 8.3 cm × 0.95 cm, occupying a volume of only 65 cm^3. While the PCB version weighed 3.178 kg, the MCM weighed only 74.4 g. These advantages were obtained at a twofold increase in cost; however, the cost can be expected to decrease as production volumes increase and manufacturing processes improve.

1.3.2 Performance

One of the most significant characteristics of the semiconductor industry is the trend towards ever-increasing clock speeds. Typical clock speeds of microprocessors around 1980 were 5-10 MHz. By 1990, clock speeds for high-end devices were in the 50-80 MHz range, and they are expected to increase to the 100-200 MHz range and beyond in this decade.

Until recently, system speeds were limited by device delays. It is perhaps surprising, then, to realize that the speed of the next generation of computers will actually be limited by the speed of light! Consider a next-generation processor running at 200 MHz. The clock cycle time is 5 ns. According to industry rules of thumb, package interconnect delays should be less than 20% of the clock cycle time, in order to guarantee a proper timing performance [8]. Thus, interconnect delays should be at most 1 ns. Then, with light traveling in a medium with $\epsilon_r = 4$, the maximum length of an interchip connection can be only 15 cm, even if we ignore the effects of loading capacitances and driver impedances. A trace on a typical PCB can easily be much longer than this. On the other hand, it is possible to fabricate a single thin-film MCM with 50 or more chips in a package measuring about 8 cm by 8 cm [8]. Clearly, multichip packages will become inevitable if such high processor speeds are to be achieved.

Multichip packaging contributes to reduced interchip signal delays in more than one way. Consider a chip-to-chip connection as shown in Fig. 1.11. The total time taken for a signal to travel from Chip 1 to Chip 2 is given by the sum of the buffer delay, time-of-flight across the interchip connection, rise-time of the received signal, and the settling time of the signal [9]. The buffer delay is a function of the loading capacitance. Since interchip connections in MCMs are finer and shorter than PCB connections, the loading capacitance is correspondingly reduced, resulting in reduced buffer delays. The time-of-flight of the signal from Chip 1 to Chip 2 is directly related to the length of the interconnection wire, and is thus reduced due to the closer chip-to-chip spacing in MCMs. Rise-time degradation depends on two factors: parasitic capacitance and inductance in the chip leads, and resistive losses in the interchip wiring. Lead parasitics are significantly reduced when single chip packages are eliminated, especially when flip-chip or overlay chip-attach technologies are used. Resistive losses, which are proportional to wire lengths, are also reduced. The settling time of the signal depends on two factors: the amount of induced noise in the interconnect, and transmission line effects such as reflections, caused by impedance mismatches along the signal path. Induced noise is proportional to wire length, and is thus reduced in MCMs. Impedance mismatches can be reduced by using controlled-impedance wiring planes sandwiched between ground planes.

1.3.3   Power consumption

The dramatic reductions in size and weight make MCMs an ideal choice for packaging in complex portable systems such as laptop computers. Another advantage which MCMs provide for portable computers is reduced power consumption. Typically, a large portion of the total power consumption of a chip is due to the off-chip drivers. These drivers are made large, since they are required to charge or discharge the large capacitive loads presented by interchip connections. The power dissipated by a CMOS off-chip driver is approximately given by the formula $P = CV^2 f$, where C is the total load capacitance of the driver, V is the voltage swing and f is the operating frequency. C is a combination of the single-chip package parasitics, chip-attach parasitics, interconnect capacitance and gate capacitances of the receiver circuits. On an MCM, the SCP parasitics are eliminated, and chip-attach parasitics are greatly reduced when flip-chip or overlay technologies are used. Interconnect capacitance is reduced due to the shorter and finer interchip connections. These factors directly contribute to reduced power consumption.

In addition, due to the reduced inductance of power and ground connections to the chips, ground bounce is significantly reduced in MCMs. Careful physical design can also reduce the levels of coupled noise. As a result, noise margins for the circuitry can be lower, and chip sets developed specifically for MCMs

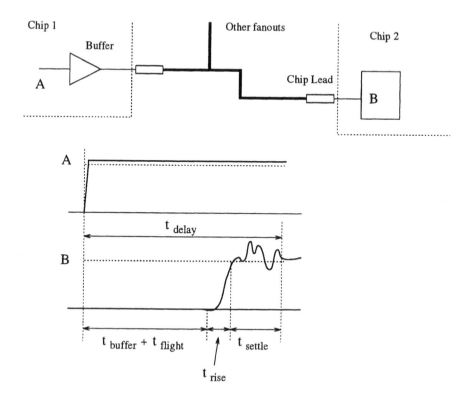

Figure 1.11 Interchip signal delay

can operate at reduced supply voltages. As the use of MCMs becomes more widespread, they will begin to have an effect earlier in the overall system design: the sizes of off-chip drivers can be reduced, since the loads to be driven are smaller. These modifications at the chip level can lead to considerable reduction in the overall power consumption. Reduced power consumption translates to prolonged battery life, reduced heatsink sizes, increased reliability and the possibility of higher operating speeds. These benefits are critical for increasing the functionality and portability of laptop or palmtop computers.

1.3.4 Reliability

By eliminating one level of the packaging hierarchy, MCMs also eliminate a large number of off-chip connections. Conventional packaging includes wire

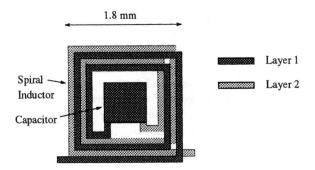

Figure 1.12 Series resonant structure in MCM substrate

bonds from the chip to the SCP, SCP leads, lead-to-board solder joints, plated-through holes (PTHs) connecting different layers in PCBs, and board connector pins. All of these are connections between dissimilar materials which are subject to failure due to mechanical and thermal stresses [6]. MCM approaches can eliminate 50-75% of these failure-prone connections, and replace many others with lower-stress joints. In the overlay interconnect process [6], for instance, the use of low stress materials such as high-ductility electroplated copper and low-modulus polymers, and the tightly controlled fabrication process for the interconnect layers, results in highly robust and reliable packages. Furthermore, the reduction in the number of intermediate packages leads to a reduction in the number of interfaces which have to be sealed, resulting in greater reliability. Robustness is essential for portable systems, which may be subjected to severe shocks, vibration and thermal stresses. The high reliability of MCMs has made them the preferred packaging technology even for space applications: an MCM for a new satellite system is described in [10].

1.3.5 Passive components and mixed-technology integration

Multichip modules offer opportunities for integrating passive components into the substrate, for further reductions in size. For example, [11] describes how to integrate spiral inductors and high-Q resonant LC circuits in MCM substrates for portable computers with built-in communication capabilities (Fig. 1.12). Designs for RF antennas integrated into the top layer of an MCM substrate which reduce loss and multipath susceptibility are also described.

Some of the advantages of MCMs, such as the size and performance improvements, can also be obtained from wafer-scale integration (WSI). However, the issue of yield continues to be a drawback for WSI. The dies on MCMs can be tested and burned-in before assembly, ensuring a very high module yield. Furthermore, MCMs offer the potential for integrating chips fabricated using different processes, such as analog and digital chips, or silicon and gallium arsenide chips, onto the same module. This feature is particularly attractive for the emerging optoelectronic and multimedia technologies.

1.4 CHALLENGES TO MCM SYSTEM DESIGNERS

Since the packaging plays a major role in determining the capabilities and limitations of a system, various packaging options should be considered from the early stages of the system design.

1.4.1 Early package analysis

Designing a VLSI system is a highly complex process in itself. The inclusion of package design in this process necessitates new approaches to handle the extra complexity. *Early analysis* is an important tool used to manage complex system designs. An early design specification is analyzed to obtain predictions of the feasibility and expected performance and cost of the system. Good models for performance and cost, which give reliable results even with incomplete input data, allow designers to know what to expect from their system, and to identify possible problem areas and limitations. Changes in the design specifications can then be made before it is too late.

An extension of the early design methodology to early *package* analysis is presented in [12]. This approach analyzes a proposed machine design in the early stages of design specification and evaluates the matching between the design and an MCM packaging technology. By considering important package-specific physical parameters, greater prediction accuracy is achieved. Another tool which evaluates MCM technology trade-offs is described in [13]. This tool assumes that a chip-level description of the system is available, rather than a gate-level description. Thus, the approach taken is an "island of integration" for package design, rather than a fully concurrent design methodology in which chips, package and system are designed simultaneously. The tool also considers thermal and cost metrics, in addition to size and electrical performance, to evaluate packaging options.

Design for packageability (DFP) is another approach which attempts to integrate packaging with the overall system design process. This concept involves taking packaging into consideration during chip design, for optimizing overall system performance. The work of [14] is focused on the impact of bonding technology (such as flip-chip, wirebonding, etc.). Design for packageability also includes the design of optimized off-chip I/O drivers for reducing the signal delay and total power dissipation in accordance with the overall system package [15].

1.4.2 Thermal design

Thermal design forms an important aspect of the package design, since it directly affects system reliability, cost, and speed. The high silicon efficiency of MCM designs and the high power dissipation of high-speed chips can push the module heat flux to several watts or tens of watts per cm². Several designs capable of handling such high flux levels are in use, or under development. IBM's thermal conduction modules use spring-loaded copper pistons to conduct heat away from the back sides of flip-chip bonded dies. For TAB or wirebonded dies, thermal vias or slugs in the substrate can be used for heat conduction. Air cooling with fans and advanced liquid cooling using microchannels in the heatsink or substrate can also be used. Figure 1.13 shows some of the available options for conducting heat away from the chips in an MCM. In addition to heat removal, thermal design also involves careful component placement to avoid "hot-spots" in the module.

1.4.3 Physical design

Physical design (placement and routing) for MCMs poses a new set of challenges, which cannot be met satisfactorily with existing VLSI and PCB design tools. A flowchart for the MCM system physical design process is shown in Fig. 1.14. The process begins with partitioning the system into chips, with the objective of maximizing performance and routability. If off-the-shelf components are used, this step is bypassed and the design process begins with the chip placement step. The size of the chip placement problem is relatively small, since MCMs rarely have over 50 chips. However, there are a number of complex constraints to consider, such as even distribution of heat over the MCM surface

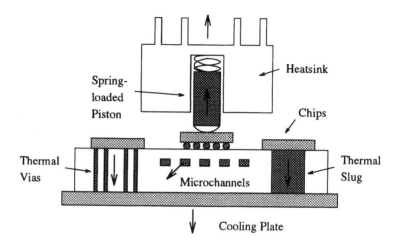

Figure 1.13 Heat removal techniques for MCM

and minimization of interchip signal delays. Traditional placement objectives such as wire length minimization are not sufficient for critical designs where wire resistance cannot be ignored. A "resistance-driven" placement approach is described in [16] to handle such situations.

In the pin redistribution step, the chip I/O pins are uniformly distributed over the MCM substrate, in order to increase routability. The next step is multilayer routing, which is usually partitioned into global two-dimensional routing and layer assignment.

In the VLSI global routing step, the primary objective is often to minimize the wire length of the global routing trees, since this reduces chip area and net delays. Thus, minimum Steiner tree algorithms have been extensively used in VLSI routing. The high wiring capacity of MCM substrates reduces the emphasis on wire length minimization. Furthermore, the resistance of an MCM interconnect is comparable to that of the driver circuit, and the inductive effects of the wire are also nonnegligible. As a result, the minimum-length tree is not necessarily the best in terms of delay. Tree-construction algorithms which take into account the resistance and/or inductance of MCM interconnects are described in [17, 18].

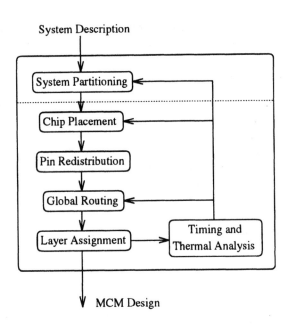

Figure 1.14 MCM physical design process

Planar routing has long been dominated by variants of grid-based maze routing algorithms. For MCMs, these algorithms have been modified to allow regular grids to be "warped" to make the most efficient use of available routing resources and optimize the routability [19]. For high-density designs, such as thin-film modules, gridded routing algorithms become very inefficient, due to the large number of grid points which must be stored in memory. Several gridless alternatives have been proposed, such as "rubber-band" routing [20], shape-based routing, and *SLICE* [21]. Rubber-band routing is a very flexible approach which easily accommodates incremental changes, supports arbitrary-angle geometries, takes advantage of unused areas to achieve evenly distributed wiring to reduce crosstalk, and allows varying-width wires. The SLICE router operates on narrow "slices" of the design, and is thus able to handle very large and complex designs efficiently. Shape-based routing stores data in terms of geometric shapes instead of on a grid, thus making more efficient use of memory resources and simplifying the routing process.

Multichip modules, especially those with laminated and ceramic substrates, may have several tens of wiring layers, as compared to just two or three for VLSI. Thus, layer assignment is an important step in the routing process. Depending on the routing approach, this may precede the layer-by-layer planar routing, or follow a two-dimensional tree generation phase. Nets are assigned to layers with several possible objectives, including minimizing the number of layers and vias, maximizing the routability on each layer, and minimizing delay and crosstalk. The number of layers may be large, and the layers may be paired into $x - y$ plane pairs. A number of MCM layer assignment algorithms have been described [22, 23, 24], and this problem continues to be an area of active research.

1.4.4   Interconnect modeling, analysis and design

The need for rapid analysis and modeling of transmission line effects and crosstalk in lossy coupled MCM interconnects has sparked a flurry of research activity recently. Rapid waveform estimation approaches based on Padé approximation techniques [25, 26, 27] are proving to be a reliable and fast approach, for situations in which accurate simulation based on convolution is prohibitively expensive. These techniques construct a reduced-order representation of the interconnect structure and compute the time-domain waveform efficiently. The approach of [27] can be used with nonlinear loads and drivers for very accurate simulation, whereas the approach of [26] is uniquely suited for sit-

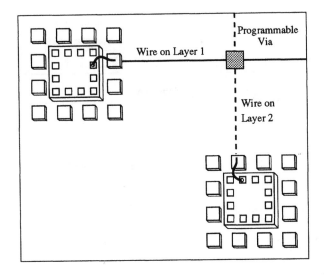

Figure 1.15 Programmable MCM interconnection

uations where the interconnect trees are incrementally modified, such as during routing. The latter approach has been used to guide an interactive/automatic MCM routing tool [28].

Manufacturing process variations can lead to variations in the electrical parameters of MCM interconnects, resulting in reduced parametric module yield. Statistical approaches have been used for yield optimization in IC design; these approaches have been extended to MCM interconnect design for maximizing module yield in [29]. The approach combines multidimensional correlated Monte-Carlo analysis and the rapid simulation technique of [25] to obtain almost a twofold increase in yield, from 44% to 82%, in one example.

Programmable MCMs can be used to reduce the turn-around time and cost of designing interconnection substrates [30]. These consist of generic substrate wafers with chip sites and two or more layers of programmable interconnections. Since the substrates can be made in large volumes, their cost can be competitive with conventional packaging. Users mount bare chips to the substrate, and complete interchip routing by setting electrically programmable switches, or using laser pulses, to connect lines on vertical routing layers to lines on horizontal layers (Fig. 1.15). The performance penalty caused by longer wire lengths and unwanted stubs can be reduced by laser trimming.

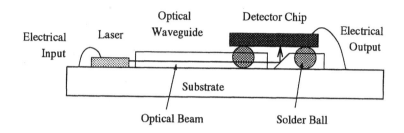

Figure 1.16 Optical interchip connections

1.4.5   Optical interconnect

Multichip modules allow the integration of different chip technologies onto the
same package. Thus, it is possible to use conventional silicon devices in conjunction with high-speed gallium arsenide optical devices for optical interchip
connections. Optical interconnects are characterized by very high bandwidth,
with negligible crosstalk between physically intersecting optical waveguides,
and negligible electromagnetic interference. Furthermore, they can have a very
high fanout without any reduction in bandwidth, unlike electrical interconnects. While optical interconnects may not replace electrical interconnects entirely, they can be used in conjunction with them to achieve faster and denser
designs, with fewer substrate layers for the electrical wiring.

A typical optoelectronic MCM setup consists of an edge emitting semiconductor laser sending optical pulses through an optical waveguide (Fig. 1.16).
The waveguide is photolithographically defined on the top layer of an MCM
substrate. The optical pulses are received by a photodetector on a flip-chip
mounted IC. This technology has been demonstrated to be compatible with
the conventional MCM manufacturing process [31]. A test structure achieved
a chip-to-chip communication rate of 1 Gb/s, which was mainly limited by
the response of the laser and photodetector – the estimated bandwidth of the
waveguide is about 45 GHz.

1.5   OVERVIEW OF THE BOOK

This book is focused on the physical design issues which arise in MCM design.
As noted earlier, one of the primary distinctions between VLSI and MCMs is

the difference in the behavior of the interconnect. The analysis of lossy coupled MCM interconnects is the topic of Chapter 2. Several recent approaches for rapid analysis of MCM interconnects are described. Delay models for resistive and inductive MCM interconnects, which can be used for performance-driven physical design, are presented.

Chapter 3 considers the problems of system partitioning and chip placement in MCMs. Problem formulations and solution techniques for timing-constrained MCM system partitioning are surveyed. A new objective function for MCM chip placement is introduced, using a delay model for lossy interconnects. Based on this objective function, a "resistance-driven" placement algorithm is developed, which reduces interconnect delays significantly in comparison with conventional placement algorithms.

The last three chapters are devoted to MCM routing. Chapter 4 begins with a description of the general multilayer MCM routing problem and describes some recent approaches used to solve it. Chapter 5 describes a different approach to multilayer routing, in which the three-dimensional problem is decomposed into two subproblems: two-dimensional routing and layer assignment. The first subproblem, two-dimensional routing or tree construction, is described, and performance-oriented algorithms for the construction of signal and clock distribution trees are presented.

Chapter 6 is devoted to the layer assignment problem in MCMs. Two different approaches to layer assignment are considered, which differ in the order in which the two-dimensional routing step and the layer assignment step are performed. An efficient algorithm for layer assignment using either approach is presented. A model for the MCM-C multilayer routing environment is developed, since this environment is different from both the VLSI and PCB routing environments. A layer assignment problem is formulated for this model, and an algorithm for solving the problem is presented.

Chapter 7 concludes the book with a discussion of open problems and directions for future research in this area.

2

ANALYSIS AND MODELING OF MCM INTERCONNECTS

2.1 INTRODUCTION

Transmission line effects in interchip interconnections on high-performance packages such as multichip modules can cause significant signal integrity problems [32]. Severe ringing caused by reflections and impedance mismatches can cause unexpectedly high delays, and crosstalk between adjacent wires can cause spurious switching. To avoid costly design iterations, accurate estimation of these phenomena during the physical design stage is essential.

Transmission line phenomena become significant at the package level because of the smaller ratio of the signal rise time to the time-of-flight of the signal across the interconnect. In on-chip interconnects, the interconnects are shorter, so transmission line effects are usually negligible. For example, the longest wire on a VLSI chip may be about 2 cm. The time-of-flight of a signal across this wire, assuming $\epsilon_r = 4$, is approximately 133 ps, which is shorter than typical on-chip signal rise times. On the other hand, the time-of-flight across a 10 cm MCM interconnect in an alumina substrate is approximately 1 ns, which is of the same order as the rise time of signals generated by off-chip drivers. Another factor which requires higher accuracy in MCM interconnect analysis is the smaller ratio of driver impedance to interconnect characteristic impedance. On-chip drivers have impedances in the kilo-ohm range, while off-chip driver impedances may be an order of magnitude smaller. Interconnect characteristic impedances are typically of the order of 100 ohms. As a result, on-chip interconnects are almost always *overdamped*, whereas MCM interconnects may be *critically damped* or *underdamped*. Overdamped systems do not exhibit large oscillations, and can be accurately modeled using first-order RC models.

Underdamped systems cannot be modeled by RC models, and require detailed simulations to predict the settling time of the output signal.

Traditional approaches to lossy transmission-line simulation suffer from high time complexity (see [33] for references to such approaches). A complex multiterminal net may take several seconds or even minutes for an accurate simulation. Clearly, the simulation step will become a bottleneck in the design process for a high-density MCM with several thousand nets.

Recently, a number of approximation techniques have been proposed for rapidly estimating the time-domain response of interconnect structures. These techniques are based on approximating the structures by systems of a smaller order, which can then be analyzed easily in the time domain. Section 2.2 reviews the most popular of these approaches, based on a technique known as *Asymptotic Waveform Evaluation* (AWE) [34]. An extension of AWE called *Complex Frequency Hopping* is described in Section 2.3. A convolution-based simulation approach which uses Padé approximations to reduce the simulation complexity is described in Section 2.4.

Section 2.5 describes a new approach for interconnect estimation, called *reciprocal expansion* (REX). Algorithms for computing the reduced-order transfer function of a lossy interconnection tree using REX are presented. Section 2.6 extends the approach to computation of capacitive crosstalk between adjacent net trees. Section 2.7 describes an extension of the REX approach which allows distributed elements to be handled directly, without resorting to lumped-element approximations. Experimental results are presented in Section 2.8.

2.2   ASYMPTOTIC WAVEFORM EVALUATION

Asymptotic Waveform Evaluation is a general technique for estimating the response of a linear (or *linearized*) system. The basic concept was first presented in [34], after which it underwent many extensions and modifications. It is now recognized that the AWE technique consists of the following steps [35]:

1. Compute the first $2n$ terms of the Maclaurin expansion of the transfer function of the system, $H(s) = m_0 + m_1 s + m_2 s^2 + \ldots + m_{2n-1} s^{2n-1} + \ldots$.

2. Convert the expansion to a strictly proper n-pole rational function, $\hat{H}(s) = \frac{P_{n-1}(s)}{Q_n(s)}$, using a Padé approximation.

3. Compute the poles and residues of $\hat{H}(s)$ to obtain the time-domain response of the reduced-order system.

These steps are described in the following sections.

2.2.1 Moment computation

The first step is the most significant: the computation of the *moments* (coefficients of the Maclaurin expansion). To understand how this is done, first consider the first-order matrix differential equation representation of a linear system:

$$\mathbf{T}\mathbf{x}(t) + \mathbf{W}\frac{d\mathbf{x}}{dt} = \mathbf{b}(t) \tag{2.1}$$

The function $\mathbf{x}(t)$ is the time-domain response of the system to input $\mathbf{b}(t)$. Rewriting Eq. (2.1) in the frequency domain, we obtain

$$(\mathbf{T} + s\mathbf{W})\mathbf{X}(s) = \mathbf{B}(s) \tag{2.2}$$

To compute the moments of the impulse response of the system, $\mathbf{B}(s)$ is replaced by $\mathbf{B}_0$, and $\mathbf{X}(s)$ is expanded in a Taylor series about $s = 0$:

$$(\mathbf{T} + s\mathbf{W})(\mathbf{X}_0 + s\mathbf{X}_1 + s^2\mathbf{X}_2 + \ldots) = \mathbf{B}_0 \tag{2.3}$$

Equating the coefficients of powers of s on the two sides of Eq. (2.3), we obtain

$$\mathbf{T}\mathbf{X}_0 = \mathbf{B}_0 \tag{2.4}$$

$$\mathbf{T}\mathbf{X}_k = -\mathbf{W}\mathbf{X}_{k-1}, k > 0 \tag{2.5}$$

Equation (2.4) is solved initially to obtain $\mathbf{X}_0$, by performing a DC analysis of the circuit, with the vector of independent sources set to the value $\mathbf{B}_0$. Using the value of $\mathbf{X}_0$, the values of the higher moments can be computed recursively using Eq. (2.5). In circuit terms, this recursive computation corresponds to performing a DC analysis, setting all independent sources to zero, replacing each capacitor by a current source equal to the product of the capacitance and the capacitor voltage computed in the previous iteration, and replacing each inductor by a voltage source equal to the inductance value times the inductor current computed in the previous iteration. Thus each moment computation corresponds to a DC steady-state solution of the same purely resistive circuit, with only the current and voltage source values changing for each computation.

2.2.2 Padé approximation

The second step in AWE is to construct a reduced-order model from the truncated series expansion of the transfer function. This is accomplished by a technique known as *Padé approximation* [36]. This is a general technique to construct a rational function of orders L and M in the numerator and denominator, respectively, which matches the first $L + M$ derivatives of a given function. For AWE, given the first $2n - 1$ moments of the system, we wish to construct an $[L/n]$ rational approximation, i.e., a rational function with a polynomial of order n in the denominator and a polynomial of order $L < n$ in the numerator. This is accomplished as follows:

$$m_0 + m_1 s + \ldots + m_{2n-1} s^{2n-1} = \frac{p_0 + p_1 s + \ldots + p_{n-1} s^{n-1}}{1 + q_1 s + q_2 s^2 + \ldots + q_n s^n} + O(s^{2n}) \quad (2.6)$$

Cross-multiplying and matching the coefficients of s on both sides of Eq. (2.6), we obtain the following sets of equations:

$$
\begin{bmatrix}
m_{L+1} & m_L & \cdots & m_{L-n+1} \\
m_{L+2} & m_{L+1} & \cdots & m_{L-n+2} \\
\vdots & \vdots & \vdots & \vdots \\
m_{L+n} & m_{L+n-1} & \cdots & m_L
\end{bmatrix}
\begin{bmatrix}
1 \\ q_1 \\ \vdots \\ q_n
\end{bmatrix}
=
\begin{bmatrix}
0 \\ 0 \\ \vdots \\ 0
\end{bmatrix}
\quad (2.7)
$$

$$
\begin{bmatrix}
m_0 & & & \\
m_1 & m_0 & & \\
m_2 & m_1 & m_0 & \\
\vdots & \vdots & \vdots & \vdots \\
m_L & m_{L-1} & m_{L-2} & \cdots & m_0
\end{bmatrix}
\begin{bmatrix}
1 \\ q_1 \\ q_2 \\ \vdots \\ q_L
\end{bmatrix}
=
\begin{bmatrix}
p_0 \\ p_1 \\ p_2 \\ \vdots \\ p_L
\end{bmatrix}
\quad (2.8)
$$

The coefficients q_i of the denominator polynomial can be computed by solving the system of n linear equalities in Eq. (2.7). The coefficients p_i can then be computed by plugging in the q_i values in Eq. (2.8).

2.2.3 Time domain response

Given the reduced-order transfer function

$$\hat{H}(s) = \frac{p_0 + p_1 s + \ldots + p_{n-1} s^{n-1}}{1 + q_1 s + q_2 s^2 + \ldots + q_n s^n} \tag{2.9}$$

the n poles of the approximate system can be computed numerically, so that the transfer function can be re-expressed as

$$\hat{H}(s) = \sum_{i=1}^{n} \frac{k_i}{s - p_i} \tag{2.10}$$

where k_i is the residue corresponding to the pole p_i. The time-domain impulse response is then given by a sum of (possibly complex) exponentials, corresponding to the inverse Laplace transform of Eq. (2.10). It should be noted that the Padé approximation technique does not guarantee that the reduced-order system is stable. Therefore, some of the computed poles may have positive real parts, which correspond to exponentially increasing functions in the time domain. Any unstable poles must be ignored when computing the time-domain response:

$$\hat{h}(t) = \sum_{Re(p_i)<0} k_i e^{p_i t} \tag{2.11}$$

The time-domain response to any arbitrary input can be computed by convolving it with the impulse response. Convolution with exponential functions can be performed recursively in linear time, as opposed to the quadratic time complexity of general convolution. For simple input functions, such as combinations of step and ramp functions, the response can be computed analytically by multiplying the Laplace transform of the input with Eq. (2.9) and extracting the poles and residues of the response waveform directly.

2.2.4 Interconnect analysis using AWE

A lossy transmission line structure can be closely approximated by an RLCG tree (Fig. 2.1.) Each unit length of the net is replaced by an RLCG section, consisting of a resistance and inductance in series, and a conductance and capacitance in shunt to ground. The accuracy of this lumped approximation depends on the maximum frequency of interest and the number of sections used. Criteria for deciding the number of sections are discussed in [37].

The transfer function of such a tree, from the source node to a particular sink node, is a $2N$th-order function of the form:

$$H(s) = \frac{a_0 + a_1 s + \cdots + a_{2N-1} s^{2N-1}}{1 + b_1 s + b_2 s^2 + \cdots + b_{2N} s^{2N}} \tag{2.12}$$

where N is the number of RLCG sections, and the a_i's and b_i's are some functions of the R, L, C and G values and the net topology. Since N may be as high as 100 or more, it is usually more convenient to find a reduced-order approximation for $H(s)$. This reduced-order approximation can be computed very efficiently in AWE, by exploiting the tree structure [25].

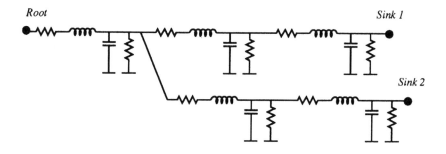

Figure 2.1   An RLCG interconnection tree

Consider the computation of the ith moment vector. For simplicity, assume that the conductance value G is zero for all sections i (this assumption can be removed easily). All capacitors are replaced by current sources and all inductors by voltage sources. By starting from the leaf nodes of the tree (i.e., by performing a reverse depth-first traversal), the current in each branch of the tree can be computed. The current entering a node u is simply the sum of all the currents entering the children of that node (Fig. 2.2). After computing the branch currents, the node voltages can be computed in a depth-first traversal of the tree, starting from the root node. The voltage at a node u is equal to the voltage at its parent node v, less the voltage drop across the resistor $R(u)$ and the voltage source corresponding to the inductor $L(u)$. Thus, the complexity of the computation of each moment is linear in the number of sections in the tree. The $RICE$ interconnect evaluation program [25] uses a generalization of the above approach, which allows it to handle deviations from a strict tree structure.

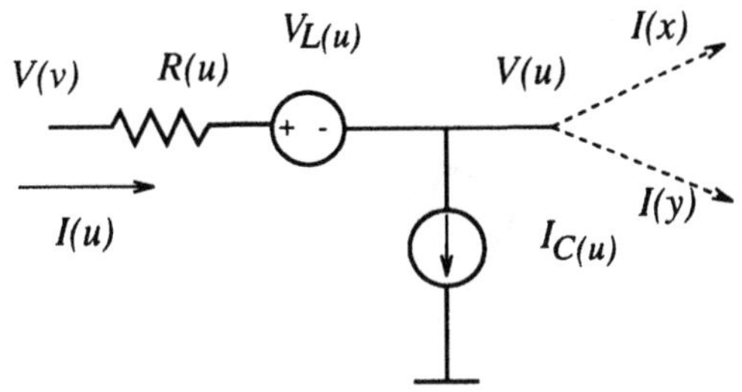

Figure 2.2   Moment computation in an RLC tree

2.2.5   Extensions to AWE

Numerous extensions to the basic concept have been developed, as described in [35]. In [38, 39], techniques are described which allow distributed elements such as transmission lines to be handled directly, without resorting to lumped circuit approximations. A variable transformation and constrained nonlinear optimization approach has been developed for extracting only stable poles from the expansion [40]. Handling of nonlinear terminations (drivers and receivers) is discussed in [41]. An improved method for generating the moments of a distributed interconnect system is presented in [42], which uses a matrix exponential computation to increase the accuracy of higher-order moments. The next section describes another approach for improving the accuracy of high-frequency poles.

2.3   COMPLEX FREQUENCY HOPPING

In AWE, a low-order approximation of the interconnect system is derived by Padé approximation from a Taylor expansion of the system transfer function about $s = 0$. The accuracy of the poles derived from this approximation depends on the distance of these poles from the origin: poles which lie outside the radius of accuracy of the Taylor expansion may be inaccurate, or may not be detected at all. Increasing the radius of accuracy of the Taylor expansion by including more terms does not necessarily improve the situation, since the higher order moments are increasingly prone to inaccuracy caused by cumulative numerical errors.

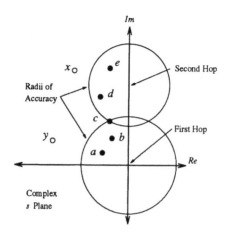

Figure 2.3 Two complex frequency hops

A solution to this problem is presented in [43], using a multi-point moment-matching technique called *Complex Frequency Hopping (CFH)*. The basic idea is to compute Taylor expansions of the circuit at a number of different (possibly complex) frequencies. The combined regions of accuracy of these "frequency hops" (Taylor expansions about some frequency) will contain more accurate poles than the single expansion about $s = 0$.

Figure 2.3 shows the effect of having two frequency hops on the reliability of computed poles. The pole marked c lies within the radii of accuracy of both the hops, so both computations yield the same value of this pole. The radius of accuracy of each hop is not known *a priori*; the computation of a common pole gives lower bounds on the radii of accuracy of the two expansions. Thus, when the common pole c is computed, it is safe to assume that poles a and b are accurate, as they are closer to the first expansion point than c is, and are thus guaranteed to lie within the radius of accuracy of the first hop. Similarly, the poles d and e are guaranteed to be accurate, whereas x and y may be inaccurate.

The CFH technique uses an efficient search technique to search the complex s−plane for all dominant poles of the network. Three constraints are used to limit the search to a finite rectangular region in the plane:

1. *Confine search to upper left-half plane.* Since the interconnect system is known to be stable, the right-half s−plane can be completely ignored.

Furthermore, all complex poles must occur in complex-conjugate pairs, so when one pole in the pair is located, the other one is known automatically.

2. *Confine search to points close to the imaginary axis.* Poles located far from the imaginary axis, i.e., poles with large negative real parts, are associated with time domain responses which decay rapidly and do not significantly affect the time-domain response. Thus the search may be confined to a region $-a \leq Re(s) \leq 0$, where a depends on the simulation time interval of interest.

3. *Confine search to points below the highest frequency.* Interconnect networks are typically low-pass networks, due to resistive losses in the wires and the capacitance of the wires and terminations. Thus, the time-domain response at the outputs of these networks do not have significant components beyond a certain maximum frequency ω_{max}, which can be estimated from the parameters of the interconnect. The CFH search can thus be limited to the region $0 \leq Im(s) \leq \omega_{max}$.

These constraints limit the search to the rectangular region shown in Fig. 2.4. Within this region, CFH uses a binary search algorithm, which begins by computing Taylor expansions at $s = 0$ and $s = j\omega_{max}$, and extracting poles. If the two expansions yield a common pole, the procedure collects all the poles within the radii of accuracy of the expansions, and computes their residues. Otherwise, another expansion is performed at the midpoint of the earlier hops, and so on, until common poles are found. The total CPU cost of the calculation depends on the number of hops required, which in turn depends on the complexity of the system, in terms of the number of dominant poles.

The CFH technique is able to extract several tens of stable poles in situations where AWE is typically limited to fewer than ten poles. The resulting waveform is of near-simulation quality, at a fraction of the CPU cost of simulation. CFH also provides an error check in the following way: if two hops are sufficiently close, they should yield a set of common poles, thus confirming the accuracy of these poles. It also provides an opportunity for trading off CPU cost for accuracy, by increasing the number of hops. This is not available in standard AWE, since simply increasing the order of the expansion does not guarantee greater accuracy.

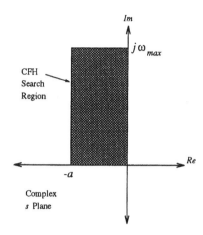

Figure 2.4 CFH search region

2.4 CONVOLUTION-BASED SIMULATION

A different approach for computing a low-order approximation of a transmission line system is presented in [27]. In this approach, the admittance and propagation function of each transmission line are approximated by low-order Padé approximants. A recursive convolution formulation is then used to compute the time-domain response of the system, which may have nonlinear terminations.

The approach begins with the Telegrapher's equations, the differential equations which express the voltage and current in a transmission line as functions of time and location along the line:

$$\frac{\partial v}{\partial x} = -\left(L\frac{\partial i}{\partial t} + Ri\right) \tag{2.13}$$

$$\frac{\partial i}{\partial x} = -\left(C\frac{\partial v}{\partial t} + Gv\right) \tag{2.14}$$

Let V_1, V_2, I_1 and I_2 denote the Laplace transforms of the voltages and currents at the input and output ports of the transmission line, respectively. The frequency domain solution of the Telegrapher's equations is given by

$$\left[\begin{array}{c} I_1 \\ I_2 \end{array}\right] = \frac{Y_0}{e^{\lambda l} - e^{-\lambda l}} \left[\begin{array}{cc} e^{\lambda l} + e^{-\lambda l} & -2 \\ -2 & e^{\lambda l} + e^{-\lambda l} \end{array}\right] \left[\begin{array}{c} V_1 \\ V_2 \end{array}\right] \tag{2.15}$$

where $Y_0 = \sqrt{\frac{G+sC}{R+sL}}$ and $\lambda = \sqrt{(G+sC)(R+sL)}$. In [44, 38], a time-domain solution is obtained by expanding the elements of the admittance matrix in Eq. 2.15 about $s = 0$. In [27], the authors argue that accuracy at high frequencies is more important for computing the transient response, since this translates to greater accuracy near $t = 0$ in the time domain. Thus, they perform the expansion about $s = \infty$. To do this, Eq. 2.15 is rearranged to obtain:

$$Y_0 V_1 - I_1 = e^{-\lambda l}(Y_0 V_2 + I_2) \tag{2.16}$$
$$Y_0 V_2 - I_2 = e^{-\lambda l}(Y_0 V_1 + I_1) \tag{2.17}$$

In the time domain, Eqs. 2.15 and 2.16 correspond to the following equations:

$$v_1(t) * h_1(t) - i_1(t) = v_2(t) * h_3(t) + i_2(t) * h_2(t) \tag{2.18}$$
$$v_2(t) * h_1(t) - i_2(t) = v_1(t) * h_3(t) + i_1(t) * h_2(t) \tag{2.19}$$

where the convolutions are performed with the following three impulse response functions:

$$h_1(t) \equiv \mathcal{L}^{-1}\{Y_0(s)\} \tag{2.20}$$
$$h_2(t) \equiv \mathcal{L}^{-1}\left\{e^{-\lambda(s)l}\right\} \tag{2.21}$$
$$h_3(t) \equiv \mathcal{L}^{-1}\left\{Y_0(s)e^{-\lambda(s)l}\right\} \tag{2.22}$$

These three inverse Laplace transforms can be derived analytically, as shown in [45]. However, to reduce computation time, these functions are approximated in [27] by sums of a small number of exponential terms in the time domain, by finding Padé approximations of $Y_0(s)$ and $e^{-\lambda(s)l}$.

2.4.1 Padé approximations of $Y_0(s)$ and $e^{-\lambda(s)l}$

The characteristic admittance $Y_0(s)$ is rewritten as

$$Y_0(s) = \sqrt{\frac{C}{L}}\sqrt{\frac{1+\frac{G}{C}y}{1+\frac{R}{L}y}} \tag{2.23}$$

where $y = 1/s$. The right hand side of Eq. 2.23 can be expanded as a Taylor series about $y = 0$, i.e., about $s = \infty$. The coefficients of the Taylor series can be computed exactly using symbolic differentiation. The first $2n$ Taylor coefficients are matched to a $[n/n]$ Padé approximant:

$$Y_0(s) \approx \sqrt{\frac{C}{L}} \frac{a_n y^n + a_{n-1} y^{n-1} + \cdots + a_1 y + 1}{b_n y^n + b_{n-1} y^{n-1} + \cdots + b_1 y + 1} \qquad (2.24)$$

$$= \sqrt{\frac{C}{L}} \frac{a_n + a_{n-1} s + \cdots + a_1 s^{n-1} + s^n}{b_n + b_{n-1} s + \cdots + b_1 s^{n-1} + s^n} \qquad (2.25)$$

The rational function is not strictly proper, since $Y_0(s)$ reaches a non-zero limit as $s \to \infty$. The poles p_i and residues q_i of the RHS of Eq. 2.24 can be computed easily, so that the inverse Laplace transform of $Y_0(s)$ can be expressed approximately as

$$h_1(t) = \mathcal{L}^{-1}\{Y_0(s)\} \approx \sqrt{\frac{C}{L}} \left(\delta(t) + \sum_{i=1}^{n} q_i e^{p_i t} \right) \qquad (2.26)$$

The exponential propagation function $e^{-\lambda(s)l}$ is approximated in a similar way, to obtain $h_2(t)$. The order of the Padé approximants for the two functions may be different. The expression for $h_3(t)$ is found by multiplying the approximants for $Y_0(s)$ and $e^{-\lambda(s)l}$ and taking the inverse transform.

2.4.2   Recursive convolution

The approximate time-domain functions $h_1(t)$, $h_2(t)$ and $h_3(t)$ consist of summations of impulses and complex exponentials (with a pure delay term in the case of h_2 and h_3). The advantage of this form is that it reduces the complexity of the convolutions in Eqs. 2.17 and 2.18 from quadratic to linear in the number of time points. Convolution with an impulse function is trivial. Consider the convolution of a function $v(t)$ with an exponential function $e^{p_i t}$, evaluated at the $(n+1)$st time point:

$$f_{n+1} = \int_0^{t_{n+1}} v(\tau) e^{p_i(t_{n+1}-\tau)} d\tau$$

$$= \int_0^{t_n} v(\tau) e^{p_i(t_{n+1}-\tau)} d\tau + \int_{t_n}^{t_{n+1}} v(\tau) e^{p_i(t_{n+1}-\tau)} d\tau$$

$$= e^{p_i(t_{n+1}-t_n)} f_n + \int_{t_n}^{t_{n+1}} v(\tau) e^{p_i(t_{n+1}-\tau)} d\tau \qquad (2.27)$$

Equation 2.26 gives a recursive expression for computing the value of the convolution at the $(n+1)$st time point from the value computed at the nth point, in constant time instead of $O(n)$ time. Thus, the overall complexity of the recursive convolution procedure is $O(n)$ as compared to $O(n^2)$ for ordinary convolution, where n is the number of time points.

2.5 RECIPROCAL EXPANSION

The AWE approach expands the transfer function $H(s)$ of a linear system as a polynomial $M(s)$. The polynomial expansion is then truncated, and used to compute a lower-order approximate transfer function, $\hat{H}(s)$.

A different approach to approximating the response of a system is to expand the *reciprocal* of the transfer function of the system:

$$\frac{1}{H(s)} = b_0 + b_1 s + b_2 s^2 + \ldots + b_{2n-1} s^{2n-1} + \ldots = B(s) \qquad (2.28)$$

This idea is the essence of the *Reciprocal Expansion* (REX) approach. For treelike structures, the coefficients of $B(s)$ can be computed efficiently, using two tree traversals in a manner not very different from the AWE computation. The REX approach also has certain desirable properties when the tree under consideration is frequently modified, by the addition or deletion of branches. This makes it useful for incremental analysis during global routing, as described in Section 2.5.4.

Assume that we are interested in finding an n-pole approximation to the transfer function $H(s)$ of an RLCG tree. The REX computation is divided into two steps, both of which are executed by efficient recursive algorithms:

Step 1: Admittance computation. In this step, for every node v in the tree, the total admittance of the subtree rooted at v is computed, in the form of a polynomial of order $2n - 1$ in s.

Step 2: Coefficient computation. In this step, the coefficients of the transfer function from the source node to the desired target node are computed. The target node can be an arbitrary node in the tree.

This approach is a generalization of the approach described in [46]. If the transfer functions to a number of different nodes in the tree are required (for instance, to all the sink nodes in a multiterminal net), step (1) is performed only once, and step (2) is executed once for each transfer function.

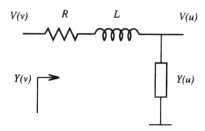

Figure 2.5 A single "RLY" section

2.5.1 Downstream admittance computation

Consider the problem of determining the *downstream admittance* $Y(v)$ of an RLCG subtree, i.e., the admittance seen "looking in" from its root node v. We shall solve this problem recursively by assuming that, for each child u of v, the downstream admittance of the subtree rooted at u is known to be $Y(u) = Y_0(u) + Y_1(u)s + \cdots + Y_{2n-1}(u)s^{2n-1}$. From Fig. 2.5, for a particular child node u, the contribution to the downstream admittance $Y(v)$ is

$$Y(v) = \frac{Y(u)}{1 + Y(u)(R(u) + sL(u))} \tag{2.29}$$

To simplify the algebraic addition of the admittance contributions of multiple children, it is convenient to express the rational function of Eq. (2.29) as a polynomial of order $2n - 1$:

$$\begin{aligned} Y &= \frac{p_0 + p_1 s + \cdots + p_{2n-1}s^{2n-1}}{1 + q_1 s + \cdots + q_{2n}s^{2n}} \\ &= Y_0 + Y_1 s + \cdots + Y_{2n-1}s^{2n-1} + H.O.T. \end{aligned} \tag{2.30}$$

The coefficients Y_i can be computed by a simple recursive formula:

$$Y_k = p_k - \sum_{i=1}^{k} q_i Y_{k-i} \tag{2.31}$$

The admittances of multiple children of a node can then be added using straightforward polynomial addition. Thus, given the downstream admittance at a node in the form of a $(2n-1)$th-order polynomial, we can find the downstream admittance at its parent node in the same form. At any leaf node v, $Y_0(v)$ and $Y_1(v)$ are known (the conductance and capacitance at that node), and the higher-order coefficients are zero. Thus, using a reverse depth-first traversal of the tree, the downstream admittance at every node in the tree can be computed as a polynomial of order $2n - 1$. The complexity of this procedure is linear in N, the number of sections, and quadratic in n, the order of the approximation (n is typically less than 10).

2.5.2 Coefficient computation

To compute the transfer function from the root to an arbitrary node x in the tree, given the looking-in admittances at every node, first consider the problem of finding the transfer function across a single "RLY" section (Fig. 2.5). In terms of the output voltage $V(u)$, the input voltage $V(v)$ can be written as

$$V(v) = V(u)[1 + Y(u)(R(u) + sL(u))] \tag{2.32}$$

Inductively assume that the node voltage at u is already known, in terms of the voltage at the target node x, as

$$\frac{V(u)}{V(x)} = b_0(u) + b_1(u)s + \cdots + b_{2n-1}(u)s^{2n-1} \tag{2.33}$$

Then $V(v)$ can be written in terms of $V(x)$ as

$$\frac{V(v)}{V(x)} = b_0(v) + b_1(v)s + \cdots + b_{2n-1}(v)s^{2n-1} + H.O.T. \tag{2.34}$$

where

$$b_k(v) = (1 + R(u)Y_0(u))b_k(u) + \sum_{i=1}^{k}(L(u)Y_{i-1}(u) + R(u)Y_i(u))\, b_{k-i}(u) \tag{2.35}$$

Starting with the values $b_0(x) = 1, b_i(x) = 0, i \geq 1$, and backtracking from the target node x to the root node r, the coefficients $b_i(r)$ can be found. We thus have:

$$\begin{aligned} \frac{V(r)}{V(x)} &= b_0(r) + b_1(r)s + \cdots + b_{2n-1}(r)s^{2n-1} \\ &= B(s) \end{aligned} \tag{2.36}$$

The complexity is again linear in N and quadratic in n.

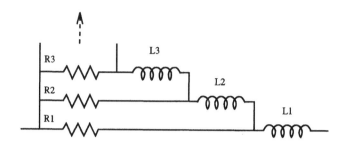

Figure 2.6 Equivalent circuit for modeling skin effect

2.5.3 Modeling of frequency-dependent effects

Although the series impedance in the RLCG sections has been assumed to consist of a resistance in series with an inductance, the same ideas can be used even when the impedance is given by a general polynomial in s. The same is true for the shunt admittance also. This generalization can be useful for modeling frequency-dependent skin effect and dielectric losses. For example, the skin effect can be modeled by a lumped RL equivalent series impedance, as shown in Fig. 2.6 [47]. The accuracy of this model increases as the number of inductors is increased. The series impedance for this model is given by the following partial fraction expansion:

$$Z(s) = sL_1 + \cfrac{1}{\cfrac{1}{R_1} + \cfrac{1}{sL_2 + \cfrac{1}{\cfrac{1}{R_2} + \cfrac{1}{sL_3 + \cfrac{1}{\cdots + \cfrac{1}{R_M}}}}}} \qquad (2.37)$$

For a given M, the R_i and L_i values can be computed as described in [47], and $Z(s)$ can be obtained as a rational function, which can then be converted to a polynomial in s.

2.5.4   Incremental admittance computation

During physical design, a routing tree may be modified frequently, by adding a new branch or deleting existing branches, while a low-order model may be used to keep track of the sink delays. In such circumstances, the downstream admittance computation does not have to be repeated for the entire tree, since the admittance values change only for those nodes "upstream" of the branch node, i.e., those nodes lying between the root and the branch node. The values can be updated by a simple procedure which involves a single backtrack from the branch node to the root. This feature is useful for on-line computation of the response during interactive or automatic routing.

When a new line is to be added to a tree at a node u, the second-order admittance looking into the line at its root can be computed analytically [46]. Let this admittance be denoted by $\Delta Y_1(u)s + \Delta Y_2(u)s^2$: this is the extra admittance added to the downstream admittance at node u. The contribution of $Y(u)$ to the downstream admittance of its parent node v is given by

$$
\begin{align}
Y_1(v) &= Y_1(u) + C(u) \tag{2.38} \\
Y_2(v) &= Y_2(u) - R(Y_1(u))^2 \tag{2.39}
\end{align}
$$

Therefore, the change in $Y(v)$ due to the addition of the new line is given by

$$
\begin{align}
\Delta Y_1(v) &= \Delta Y_1(u) \tag{2.40} \\
\Delta Y_2(v) &= \Delta Y_2(v) - 2RY_1(u)\Delta Y_1(u) - R(\Delta Y_1(u))^2 \tag{2.41}
\end{align}
$$

Equations (2.39) and (2.40) can be used to recursively increment the downstream admittances of all nodes upstream of u. The procedure for updating admittances when an existing branch is deleted is similar: only the sign of $\Delta Y(u)$ is reversed.

2.6   EXTENSION TO COUPLED TREES

An important phenomenon responsible for signal degradation in high-density interconnects is the *crosstalk* or coupling between long adjacent wires. Simulation of crosstalk waveforms is a difficult task, and is currently not supported by the popular circuit simulation program Spice (Version 3E).

The procedures described in the previous section for isolated trees can be extended to a pair of capacitively coupled trees, when the region of coupling is

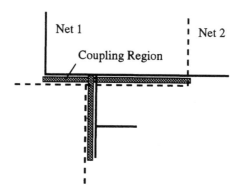

Figure 2.7 Capacitively coupled trees

itself in the form of a tree (see Fig. 2.7). Although the coupling introduces loops in the circuit, the simulation algorithms retain their linear complexity.

The crosstalk computation procedure also proceeds in two steps: admittance computation followed by coefficient computation. The result will be a pair of coupled equations of the form:

$$\begin{aligned}
V_r(s) &= A(s)V_0(s) + B(s)U_0 s \\
U_r(s) &= C(s)V_0(s) + D(s)U_0(s)
\end{aligned} \tag{2.42}$$

where V_r and U_r are the voltages at the roots of the two trees, V_0 and U_0 are the voltages at target points on the two trees, and $A, B, C,$ and D are polynomials of order $2n - 1$ in s. The crosstalk waveform at V_0 due to a pulse at U_r can be computed by setting $V_r = 0$ and solving for V_0:

$$\frac{V_0(s)}{U_r(s)} = \frac{B(s)}{B(s)C(s) - A(s)D(s)} \tag{2.43}$$

2.6.1   Generalization of admittance computation

Let the two coupled trees be denoted by T_1 and T_2, and the region in which the coupling occurs be denoted by T_c. We assume that T_c is also a tree. Every edge in T_c corresponds to two RLCG sections, one in T_1 and one in T_2, and every node in T_c corresponds to a pair of coupled nodes, one each in T_1 and

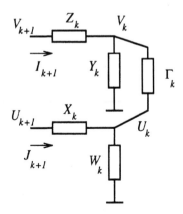

Figure 2.8 Coupled RLCG sections

T_2. The coupled nodes are connected by a *coupling admittance* γ_K, which is typically a capacitance. For the single tree case, we recursively computed the admittance $Y(v)$ at a node v, which gave us the ratio between the current flowing out of node v and the voltage at node v. For the case of coupled trees, the computation is generalized as follows. For a pair of coupled nodes, we define three quantities, W_u, Y_v and Γ_{uv}:

$$\begin{aligned} I_v &= Y_v V_v + \Gamma_{uv}(V_v - U_u) \\ J_u &= W_u U_u + \Gamma_{uv}(U_u - V_v) \end{aligned} \tag{2.44}$$

where U_u, V_v and J_u, I_v are the voltages at, and currents flowing out of, nodes u and v, respectively.

Given the admittances at a pair of coupled nodes u_k and v_k, we wish to compute the corresponding quantities at their parent nodes u_{k+1} and v_{k+1} (see Fig. 2.8), so as to obtain a recursive procedure similar to that for the single-tree case. Writing equations for the input currents I_{k+1} and J_{k+1}, we have

$$\begin{aligned} I_{k+1}D_k &= [Y_k(1 + (W_k + \Gamma_k)X_k) + \Gamma_k(1 + W_k X_k)]V_{k+1} - \Gamma_k U_{k+1} \\ J_{k+1}D_k &= [W_k(1 + (Y_k + \Gamma_k)Z_k) + \Gamma_k(1 + Y_k Z_k)]U_{k+1} - \Gamma_k V_{k+1} \end{aligned} \tag{2.45}$$

where

$$\begin{aligned} D_k &= 1 + W_k X_k + Z_k Y_k + \Gamma_k(X_k + Z_k) \\ &\quad + \Gamma_k X_k Z_k(W_k + Y_k) + W_k X_k Y_k Z_k \end{aligned} \tag{2.46}$$

Comparing (2.44) with the form of (2.43), we obtain:

$$Y_{k+1} = \frac{Y_k(1 + (W_k + \Gamma_k)X_k) + \Gamma_k W_k X_k}{D_k} \tag{2.47}$$

$$W_{k+1} = \frac{W_k(1 + (Y_k + \Gamma_k)Z_k) + \Gamma_k Y_k Z_k}{D_k} \tag{2.48}$$

$$\Gamma_{k+1} = \frac{\Gamma_k}{D_k} \tag{2.49}$$

To compute the admittances at all nodes in the two trees, the admittances of those parts of the two trees which extend beyond the leaves of the coupling region are computed first, using the recursion described for the single-tree case. This gives us values for Y and W at all the leaves of the coupling region T_c. The value of Γ at each leaf of T_c is simply the coupling admittance γ. Equations (2.47)–(2.48) are then used to recursively compute the admittances at all nodes of T_c (which are pairs of coupled nodes in T_1 and T_2). The root node of T_c is made to coincide with the root nodes of both trees T_1 and T_2, by conceptually adding "dummy" coupling admittances between uncoupled RLCG sections, and/or adding extra dummy segments to one of the trees, if necessary.

2.6.2   Coupled coefficient computation

The coefficients of the polynomials A, B, C and D in Eq. (2.41) are computed by backtracking from the target nodes of T_1 and T_2. The backtracks on the two trees proceed separately, until they both hit a leaf l of T_c. (In order to ensure that both of the backtrack procedures hit the *same* leaf of T_c, only one of the two target nodes can be chosen arbitrarily.) Then, we backtrack from the leaf of T_c to its root (which is also the root of T_1 and T_2), applying the following recursions:

$$V_{k+1} = V_k(1 + Z_k(Y_k + \Gamma_k)) - Z_k\Gamma_k U_k \tag{2.50}$$

$$U_{k+1} = U_k(1 + X_k(W_k + \Gamma_k)) - X_k\Gamma_k V_k \tag{2.51}$$

At the leaf l of T_c, the initial separate backtrack computations give us the starting points for the recursion, expressing U_l/U_o and V_l/V_o as polynomials in s. The polynomials are updated at each node in T_c, and at the root, we obtain the final values of the coefficients of A, B, C and D.

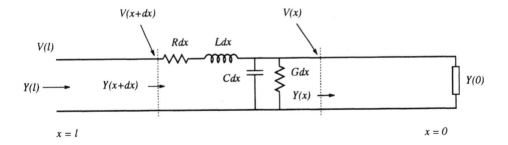

Figure 2.9 Distributed-parameter transmission line

2.7 DIRECT MODELING OF DISTRIBUTED ELEMENTS

The approach taken for lossy transmission line simulation, as described above, has involved approximating distributed-parameter systems by a large number of lumped RLCG sections, followed by an approximation of the transfer function of the lumped element system by a lower-order function. This section describes how the first approximation can be avoided by modeling the distributed-parameter elements directly. This approach can be used to increase efficiency when the transmission lines are uniform.

The distributed-parameter analysis is presented in two sections: in the first part, we describe how a second-order model can be constructed rapidly for an interconnection tree. A second-order model is useful for quick delay analysis during routing, as described in Section 2.9 and Chapter 5. The second part describes an approach for constructing a general higher-order model.

2.7.1 Second-order distributed model

Figure 2.9 shows a distributed-parameter transmission line. The quantities R, L, C and G are all per unit length. Suppose the line is terminated in an admittance of $Y(0) = Y_1(0)s + Y_2(0)s^2$ at the far end, i.e., at $x = 0$. We wish to determine the looking-in admittance at the near end $(x = l)$ and the voltage transfer ratio, $\frac{V(l)}{V(0)}$. Consider an infinitesimal section of the line, at position x as shown in the figure. In this analysis, we ignore the shunt conductance, (i.e., assume $G = 0$), since it is usually very small in real conductors and does not significantly affect the delay computation. The expression for the admittance at $(x + dx)$ is as follows:

$$Y(x + dx) = \frac{Y(x) + sC\,dx}{1 + dx(R + sL)(Y(x) + sC\,dx)} \qquad (2.52)$$

Assuming that $Y(x)$ is given by a polynomial in s, $Y(x) = Y_1(x)s + Y_2(x)s^2 + \cdots$, the rational function on the right-hand side of Eq. (2.52) can be expanded as a Maclaurin series:

$$
\begin{aligned}
Y(x + dx) &= Y_1(x + dx)s + Y_2(x + dx)s^2 + \cdots & (2.53)\\
&= (Y_1(x) + C\,dx)s \\
&\quad + (Y_2(x) - R\,dx(Y_1(x) + C\,dx)^2)s^2 + \cdots & (2.54)
\end{aligned}
$$

Equating the coefficients of the first two powers of s on the right-hand sides of Eqs. (2.52) and (2.53) gives the following differential equations for $Y_1(x)$ and $Y_2(x)$:

$$\frac{dY_1(x)}{dx} = C \qquad (2.55)$$

$$\frac{dY_2(x)}{dx} = -R(Y_1(x))^2 \qquad (2.56)$$

Solving Eq. (2.54) yields

$$Y_1(x) = Y_1(0) + Cx \qquad (2.57)$$

Substituting Eq. (2.57) in Eq. (2.55) and integrating, we have

$$Y_2(x) = Y_2(0) - Rx(\frac{(Cx)^2}{3} + CxY_1(0) + (Y_1(0))^2) \qquad (2.58)$$

Equations (2.57) and (2.58) allow us to compute the second-order looking-in admittance $Y(l)$ in terms of the admittance $Y(0)$. They are thus analogous to Eqs. (2.29) and (2.29) for the lumped-element situation.

We next consider the second-order voltage-transfer coefficients. Suppose the voltage at $x = 0$ is given by $(1 + b_1(0)s + b_2(0)s^2 + \cdots)V_{out}$, where V_{out} is the voltage at the sink node of interest. We wish to express $V(l)$ in a similar form, i.e., as $V(l) = (1 + b_1(l)s + b_2(l)s^2 + \cdots)V_{out}$. To do this, we develop differential equations for $b_1(x)$ and $b_2(x)$. Consider the voltage at $(x + dx)$:

$$V(x + dx) = V(x)(1 + dx\,(R + sL)((Y_1(x) + C\,dx)s + Y_2(x)s^2)) \qquad (2.59)$$

This can be rewritten as

$$
\begin{aligned}
\frac{dV(x)}{dx} &= (1 + b_1(x)s + b_2(x)s^2)(RY_1(x)s + (RY_2(x) + LY_1(x))s^2) & (2.60)\\
&= RY_1(x)s + (RY_1(x)b_1(x) + RY_2(x) + LY_1(x))s^2 & (2.61)
\end{aligned}
$$

This gives us the following differential equations for $b_1(x)$ and $b_2(x)$:

$$\frac{db_1(x)}{dx} = RY_1(x) = RCx + RY_1(0) \tag{2.62}$$

$$\frac{db_2(x)}{dx} = RY_1(x)b_1(x) + RY_2(x) + LY_1(x) \tag{2.63}$$

Solving Eq. (2.61) yields

$$b_1(x) = b_1(0) + \frac{(Rx)(Cx)}{2} + RY_1(0)x \tag{2.64}$$

Substituting Eqs. (2.57), (2.58) and (2.64) in Eq. (2.62) and integrating, we have

$$b_2(x) = b_2(0) + x(R(Y_2(0) + Y_1(0)b_1(0)) + LY_1(0))$$
$$+ \frac{x^2}{2}(RCb_1(0) + LC) + \frac{x^3}{6}R^2CY_1(0) + \frac{x^4}{24}R^2C^2 \tag{2.65}$$

Equations (2.64) and (2.64) are analogous to Eqs. (2.29) and (2.35), and can be used to compute the second-order coefficients of $V(l)$.

2.7.2 Higher-order modeling of distributed elements

Two approaches to computing the admittance and voltage-transfer polynomials to arbitrary degrees are described in this section. The first approach assumes that $G = 0$ and $Y_0(0) = 0$, and is an extension of the method described in the previous section. The second approach is appropriate for general lines with nonzero G and $Y_0(0)$.

Modeling with $G = 0$

Consider the distributed line of Fig. 2.9. With $G = 0$, the admittance at a point $(x + dx)$ can be written as

$$Y(x + dx) = \frac{Y(x) + sC\,dx}{1 + dx\,(R + sL)(Y(x) + sC\,dx)} \tag{2.66}$$

Subtracting $Y(x)$ from both sides of Eq. (2.66), we have

$$Y(x + dx) - Y(x) = \frac{sC\,dx - Y(x)(R + sL)\,dx(sC\,dx + Y(x))}{1 + (R + sL)\,dx(sC\,dx + Y(x))} \tag{2.67}$$

In the limit as $dx \to 0$, we obtain the following differential equation:

$$\frac{dY(x)}{dx} = sC - Y^2(x)(R + sL) \qquad (2.68)$$

Suppose that $Y(x)$ is a polynomial of order n in s, with the constant term = 0, i.e.,

$$Y(x) = Y_1(x)s + Y_2(x)s^2 + \cdots + Y_n(x)s^n \qquad (2.69)$$

From Eqs. (2.68) and (2.69), we obtain the following differential equations for the coefficients $Y_k(x)$:

$$\frac{dY_1(x)}{dx} = C \qquad (2.70)$$

$$\frac{dY_k(x)}{dx} = -R\sum_{i=1}^{k-1} Y_i(x)Y_{k-i}(x) - L\sum_{i=1}^{k-2} Y_i(x)Y_{k-1-i}(x) \qquad (2.71)$$

Lemma 2.1 *Each coefficient $Y_i(x)$ is a polynomial of order $2i - 1$ in x.*

Proof: By induction. Integrating Eq. (2.69), we obtain $Y_1(x) = Y_1(0) + Cx$. Thus the statement is true for $i = 1$. Suppose it is true for all $i < k$. Then, from Eq. (2.70), the derivative of $Y_k(x)$ is a polynomial in x, of order $(2i - 1) + (2(k - i) - 1) = 2k - 2$. Therefore, $Y_k(x)$ is a polynomial of order $2k - 1$ in x.

The polynomials $Y_k(x)$ can be computed recursively, beginning with $Y_1(x)$, by integrating Eq. (2.70). Since the integrand is a polynomial, evaluation of the integral is trivial and can be performed as either a definite or an indefinite integral. Thus, we have a recursive technique for computing the coefficients $Y_k(l), k = 1, \ldots, n$.

We now consider the line voltage at point $x + dx$:

$$V(x + dx) = V(x)\left[1 + (R + sL)\, dx(sC\, dx + Y(x))\right] \qquad (2.72)$$

where

$$V(x) = V_{out} \sum_{i=0}^{n} b_i(x)s^i \qquad (2.73)$$

From the above equations, we obtain the following differential equations for the coefficients $b_k(x), k = 1, \ldots, n$ (note that $b_0(x) \equiv 1$):

$$\frac{db_k(x)}{dx} = \sum_{i=0}^{k-1} b_i(x)\left[RY_{k-i}(x) + LY_{k-1-i}(x)\right] \qquad (2.74)$$

Lemma 2.2 *Each coefficient $b_i(x)$ is a polynomial of order $2i$ in x.*

The proof is very similar to that for Lemma 2.1. Thus Eq. (2.74) gives us a recursion for computing the polynomials $b_k(x), k = 1, \ldots, n$, starting from $k = 1$, using the previously computed polynomials $Y_i(x)$.

To summarize, given the following input:

1. A transmission line of length l, with constant parameters R, L, C ($G = 0$),

2. A terminating admittance at the far end, $Y(0) = \sum_{i=1}^{n} Y_i(0)s^i$, and

3. A polynomial $1 + \sum_{i=1}^{n} b_i(0)s^i$ expressing the far-end voltage in terms of the desired target node voltage V_{out},

the method described above allows us to compute

1. The looking-in admittance at the near end, $Y(l) = \sum_{i=1}^{n} Y_i(l)s^i$, and

2. A polynomial $1 + \sum_{i=1}^{n} b_i(l)s^i$ expressing the near-end voltage in terms of V_{out}.

Thus, the response of an arbitrary tree structure consisting of transmission lines can be analyzed without resorting to lumped element approximations. Each transmission line must be uniform, but any two lines may have different parameters.

Modeling with $G \neq 0$

When the conductance G of the transmission line is not negligible, or when terminating resistors are used in the interconnect, the solutions to the differential equations (2.70) and (2.74) are not polynomials, and are considerably more difficult to compute. This situation can be handled by a different approach.

Consider a uniform transmission line terminated by an impedance Y_2, as shown in Fig. 2.10. The solution for the Telegrapher's equations in this line can be

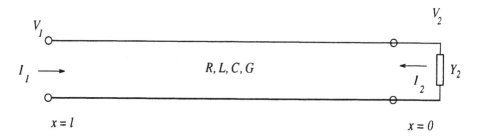

Figure 2.10   Uniform transmission line

expressed as:

$$\begin{bmatrix} I_1 \\ I_2 \end{bmatrix} = \frac{Y_0}{e^{\lambda l} - e^{-\lambda l}} \begin{bmatrix} e^{\lambda l} + e^{-\lambda l} & -2 \\ -2 & e^{\lambda l} + e^{-\lambda l} \end{bmatrix} \begin{bmatrix} V_1 \\ V_2 \end{bmatrix} \tag{2.75}$$

where $Y_0 = \sqrt{\frac{G+sC}{R+sL}}$ and $\lambda = \sqrt{(G+sC)(R+sL)}$. Substituting $I_2 = -Y_2 V_2$ in Eq. (2.75) and solving for I_1/V_1 and V_1/V_2, we obtain

$$\frac{I_1}{V_1} = Y_1 = Y_0 \frac{(e^{\lambda l} - e^{-\lambda l}) + (e^{\lambda l} + e^{-\lambda l})Y_2/Y_0}{(e^{\lambda l} + e^{-\lambda l}) + (e^{\lambda l} - e^{-\lambda l})Y_2/Y_0} \tag{2.76}$$

$$\frac{V_1}{V_2} = \frac{(e^{\lambda l} + e^{-\lambda l})}{2} + \frac{(e^{\lambda l} - e^{-\lambda l})}{2} \frac{Y_2}{Y_0} \tag{2.77}$$

The quantities Y_0, $(e^{\lambda l} + e^{-\lambda l})$ and $(e^{\lambda l} - e^{-\lambda l})$ can be expanded in Taylor series expansions using symbolic differentiations, as described in [27]. These expansions give us the desired polynomials for looking-in admittance and voltage transfer ratio.

2.7.3   Extraction of pure delay

Due to the finite time taken for an electromagnetic wave to travel from one end of a transmission line to another, the voltage transfer function of the line contains a pure delay term, represented in the frequency domain by e^{-sT}, where T is the delay, given by $l\sqrt{LC}$. When the transfer function is approximated using the techniques described above, the delay cannot be approximated well, and the time-domain response of the system for $0 \le t < T$ contains some spurious oscillations, which are not present in the actual response. To avoid these oscillations, it is possible to extract the pure delay term from the voltage transfer ratio polynomial beforehand, as follows: suppose the computed polynomial is

$$\frac{V_{in}}{V_{out}} = \sum_{i=0}^{n} b_i s^i \tag{2.78}$$

which we would like to convert to the form

$$\frac{V_{in}}{V_{out}} = e^{sT} \sum_{i=0}^{n} \beta_i s^i \tag{2.79}$$

The coefficients β_i can be computed as follows:

$$\beta_k = \sum_{j=0}^{k} b_j \frac{(-T)^{k-j}}{(k-j)!}, \quad k = 0, \ldots, n \tag{2.80}$$

The pure delay terms can be added up when multiplying the transfer ratio polynomials of cascaded lines. The time-domain response of the lower-order system can be computed by ignoring the delay while extracting the poles and residues, and then time-shifting the computed response of the system to the right by T seconds.

2.8   EXPERIMENTAL RESULTS

The accuracy of the REX approach described in the previous sections was verified by comparing the computed step response waveforms for a large number of multiterminal nets, with accurate lossy transmission line simulations using Spice (Version 3E), and approximate responses using the AWE method. The number of lumped RLCG sections used for the nets in REX and AWE was typically between 50 and 100. The line parameters and driver impedance were all varied over a wide range, and uniformly accurate results were obtained in all cases.

Figures 2.11 and 2.12 show the responses obtained using a seven-pole model for a six-terminal net, with different driver impedances (the transmission line impedance was 50 Ω). The REX approach took less than 0.3 sec for each simulation, as compared to more than 220 sec for each Spice simulation. The high

V

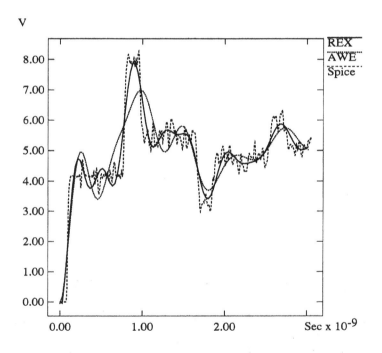

Figure 2.11 Step response for $R_d = 10\,\Omega$

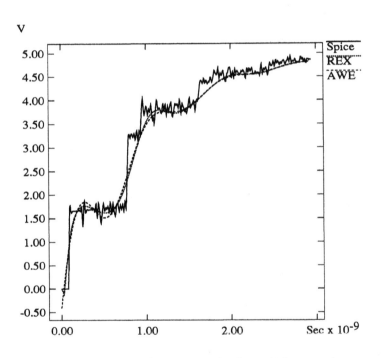

Figure 2.12 Step response for $R_d = 100 \, \Omega$

Table 2.1 Electrical parameter values used in the comparison

Parameter	Symbol	Value
Driver Resistance	R_d	10Ω
Sink Capacitance	C_s	0.01 pF
Resistance/Unit Length	R	$0.2\ \Omega$
Inductance/Unit Length	L	0.33 nH
Capacitance/Unit Length	C	0.133 pF
Conductance/Unit Length	G	0.0
Characteristic Impedance	$Z_o = \sqrt{L/C}$	$50\ \Omega$

accuracy of the waveform in these two figures is typical of all the experiments we have conducted.

Figure 2.11 also shows the waveform obtained using the AWE approach – the accuracy of the waveform is not as high as that of the REX waveform. In general, it was observed that for underdamped situations (when the driver resistance is less than the line impedance), the REX approach is more accurate than AWE, and for the matched and overdamped situations, both approaches are equally accurate on the average. This result is significant because it is the underdamped situation which is difficult to simulate accurately, requiring a large number of poles (overdamped lines can be approximated well even by first- or second-order models).

The accuracy of REX was compared experimentally with that of the AWE approach, on a set of 10 different multiterminal nets, with a total of 34 sink nodes. The electrical parameter values used in the simulations are shown in Table 2.1. The nets were all underdamped, with a driver resistance of 10 Ω and a line impedance of 50 Ω. The number of unstable poles generated using REX was less than or equal to the corresponding number using AWE in all but five of the 34 cases. On the average, AWE generated 21% more unstable poles than REX. The residues associated with unstable poles were significant (> 0.1) in seven cases using AWE, whereas significant residues occurred in only two cases using REX. The average unstable residue using AWE was three times as large as the average unstable residue using REX.

Figure 2.13 shows the crosstalk waveform caused by capacitive coupling between two adjacent lines with six RLCG sections each. A five-pole approximation was

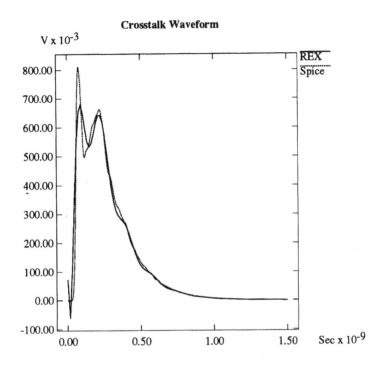

Figure 2.13 Crosstalk waveform for two coupled lines

used for the computed waveform. The Spice waveform was found by simulating the lumped RLC circuit, since coupled transmission line simulation is not handled by Spice 3E.

Figures 2.14 and 2.15 show two examples of the effect of extraction of the pure delay term. The spurious oscillations in the region $0 \leq t < T_d$ are eliminated, and the overall accuracy of the waveform is improved.

Accuracy improvement due to direct modeling of distributed elements is illustrated in Figs. 2.16 and 2.17. In Fig. 2.16, a transmission line was modeled using only two lumped RLC sections. Thus, the approximate transfer function using the lumped approach can have at most four poles. The distributed model, on the other hand, can have any number of poles. In the figure, the lumped model has four poles and the distributed model has six poles. Figure 2.17 compares the lumped response of another line, which was modeled using five lumped sections, with the distributed model response.

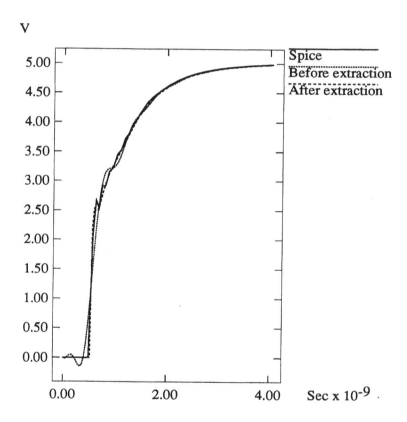

Figure 2.14   Pure delay extraction for a critically damped line

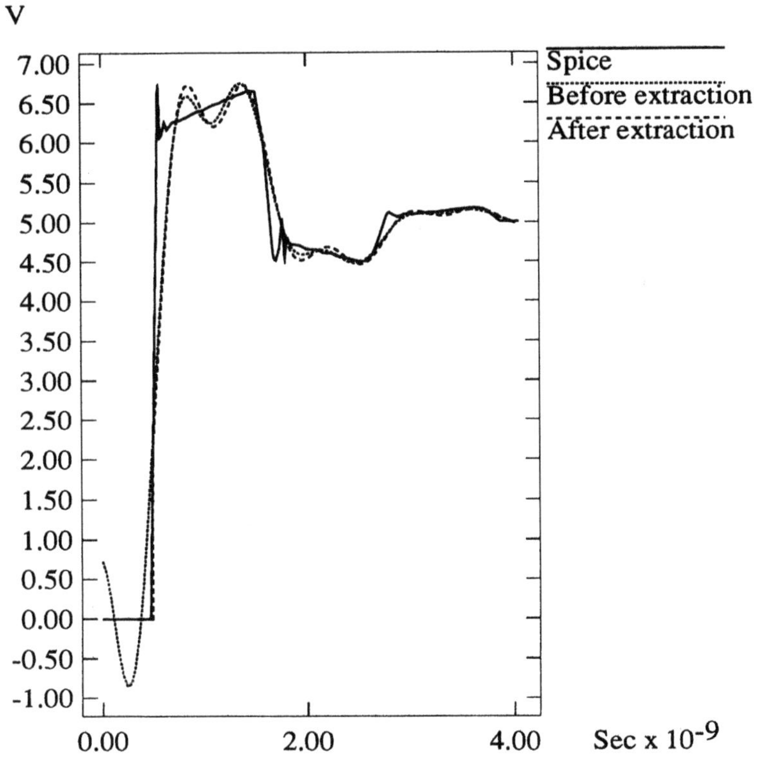

Figure 2.15 Pure delay extraction for an underdamped line

V

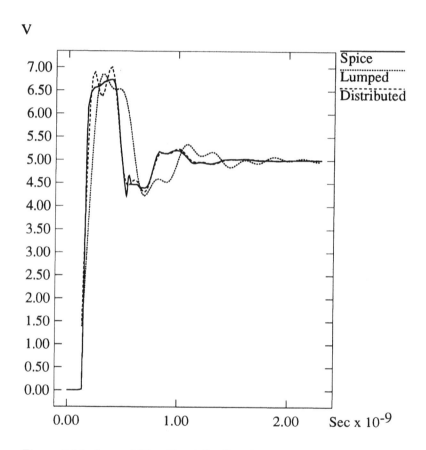

Figure 2.16   Lumped ($N = 2$) and distributed model responses

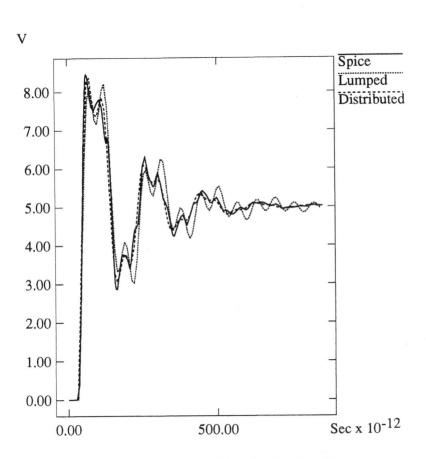

Figure 2.17 Five-section lumped model vs. distributed model

2.9   DELAY MODELING OF MCM INTERCONNECTS

VLSI interconnect delay modeling is conventionally done in one of two ways, depending on the degree of accuracy required:

Lumped capacitor model: In this model, the interconnect is treated as a single lumped capacitor, with capacitance C equal to the total distributed capacitance of the interconnect. Interconnect delay is then estimated as $R_d C$, where R_d is the output impedance of the driver circuit.

RC tree model: The lumped capacitor model ignores the resistance of the interconnection wire. As feature sizes decrease, wire resistance increases, invalidating this approximation. The resistance of the interconnect can be taken into account by approximating the distributed RC structure using a lumped RC tree. Section 2.9.1 describes two useful first-order RC delay models.

In packages, even the RC tree model is inadequate to accurately model delay, because the driver resistance is of the same order or smaller than the characteristic impedance of the interconnect, resulting in underdamped transmission line systems. The RC models cannot model higher-order underdamped systems. Increasing the order of the delay model from the first-order to any arbitrary order cannot improve the accuracy significantly, because the voltage at any node in an RC circuit with zero initial conditions is always monotonically rising (or falling), whereas there may be overshoots and ringing in underdamped transmission line systems. A detailed delay model for a nonlinear driver and a single transmission line is presented in [48]; however, this model does not handle the case of tree-structured interconnects.

A second-order RLC model of the interconnect can provide significant improvement over the accuracy of a first-order RC delay model, since RLC trees and second-order transfer functions may have oscillatory output voltages, and are thus able to predict increased delays due to settling times of interconnect output waveforms. Section 2.9.2 presents a second-order RLC delay model based on the REX approach.

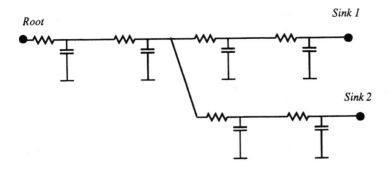

Figure 2.18 A lumped RC tree

2.9.1 First-order delay models

One of the most popular first-order RC delay models is the *Elmore time constant* [49]. Consider a multiterminal interconnect structure modeled as a lumped RC tree (Fig. 2.18). The Elmore delay from the root or driver node d to a particular sink node s_i is computed as

$$t_{Elmore}(T) = \sum_{k \in p(d,s_i)} C'_k R_{p(d,k)} \tag{2.81}$$

where $p(d, s_i)$ is the *main path* (path from d to s_i); C'_k is the total *off-path* capacitance from node k to ground, i.e., the capacitance at node k plus the total capacitances of all subtrees rooted at k (except the main path itself); and $R_{p(d,k)}$ is the resistance of the path from d to k. This expression corresponds to the first moment of the transfer function from d to s_i, and is a very useful delay prediction tool.

Another expression which can be used for delay estimation in an RC tree is the *dominant time constant* delay model. The expression for this quantity is

$$t_P(T) = \sum_{k \in T} C_k R_{p(d,k)} \tag{2.82}$$

where C_k is the value of the capacitance at node k. The quantity $t_P(T)$ is equal to the negative sum of the reciprocals of all the poles of the system. This expression is used frequently to estimate the bandwidth of multistage amplifiers. For any tree T, Eq. (2.82) gives a single delay figure, while the Elmore delay model gives one delay value for each sink node. It is easy to show that $t_P(T) \geq t_{Elmore}(T)$ for any tree T [50]. The dominant time constant model is thus more pessimistic than the Elmore model. However, it is also easier to use during physical design since it gives a single delay figure for all terminals of a multiterminal net, and is easier to compute.

2.9.2 Second-order RLC delay model

Using a lumped RLC tree model for an interconnect structure, and the REX technique described earlier, it is easy to find a second-order approximation for the transfer function from the driver to a particular sink node:

$$\hat{H}(s) = \frac{a_0}{1 + b_1 s + b_2 s^2} \tag{2.83}$$

If all shunt conductances in the tree are negligible (as is usually the case), $a_0 = 1$. Given the values of b_1 and b_2, it is possible to estimate the delay from d to the sink node with a step input at d. We first define the following quantities:

$$K = \frac{4b_2}{b_1^2} \tag{2.84}$$

$$V_{th} = (1 - 1/e)V_{final} \tag{2.85}$$

$$t_{norm} = \frac{t_d}{b_1} \tag{2.86}$$

where t_d is the time taken for the step response to settle to a value greater than the threshold value of $(1 - 1/e)$ times the final voltage.

Depending on the value of K, we identify three possible cases:

Monotone Case $(K < 1)$. For these values of K, the system step response is the sum of two real exponentials, and is thus monotone rising (Fig. 2.19). An analytical expression cannot be found for this situation, but $t_{norm} = 1$ is a reasonable approximation. An accurate empirical expression is described below.

Small Oscillations $(1 \leq K \leq 4\pi^2 + 1)$ When K is greater than unity, the system response is oscillatory (Fig. 2.20). However, the largest undershoot of the response does not fall below V_{th} until K reaches the value of $4\pi^2 + 1$. Therefore, for this range of values of K, the delay is still given by the first time point at which the output voltage crosses V_{th}.

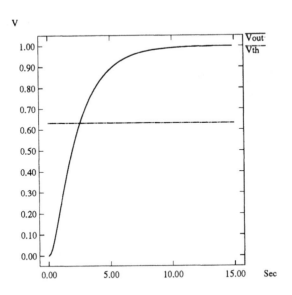

Figure 2.19   Monotone step response

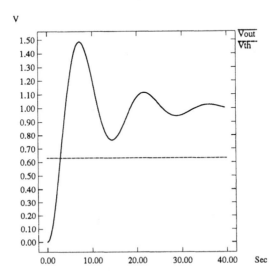

Figure 2.20   Oscillatory step response

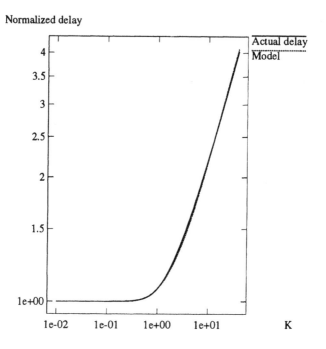

Figure 2.21   Empirical delay model for $K < 4\pi^2 + 1$

An empirical formula which closely fits the actual delay values over the range $0 \leq K \leq 4\pi^2 + 1$ is the following:

$$t_{norm} = \frac{1}{1 - ae^{-bK^{-0.68}}} \qquad (2.87)$$

where $a = 0.93745, b = 2.62321$. This expression estimates t_{norm} with less than 2% error over the range $0 \leq K \leq 4\pi^2 + 1$, as shown in Fig. 2.21.

Large Oscillations $(K > 4\pi^2 + 1)$ When K crosses the value $4\pi^2 + 1$, the first undershoot falls below V_{th}, and the delay must be computed as the last time point at which the response crosses V_{th}. The actual delay is discontinuous, as illustrated in Fig. 2.23: the nth discontinuity corresponds to the value of K for which the nth undershoot falls below V_{th}. A simple approximation for this situation is to use the *lower envelope* of the output waveform (Fig. 2.22). The lower envelope is given by the following

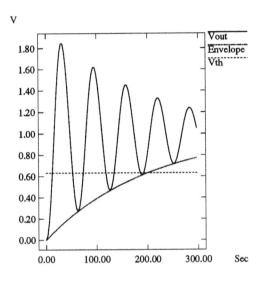

Figure 2.22   Step response with large oscillations

expression:

$$v_{LE}(t) = 1 - \sqrt{\frac{K}{K-1}} e^{-\frac{2}{K}\frac{t}{b_1}} \tag{2.88}$$

The time point at which v_{LE} crosses V_{th} is given approximately by $t_{norm} = K/2$. As shown in Fig. 2.23, this results in a pessimistic estimate of the delay. However, this is not unreasonable, because if the interconnect is severely underdamped, there may be several large-amplitude oscillatory modes in the response, of which the second-order model captures only the most prominent one. The other oscillations may increase the actual delay.

Thus the overall delay model is given by the following expressions (see Fig. 2.24):

$$t_d = \begin{cases} \dfrac{1}{1 - 0.93745e^{-2.62321K^{-0.68}}}b_1 & K \le 4\pi^2 + 1 \\[2mm] \dfrac{K}{2}b_1 & K > 4\pi^2 + 1 \end{cases} \tag{2.89}$$

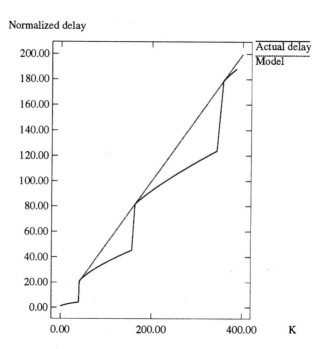

Figure 2.23   Delay model for large K values

Normalized delay

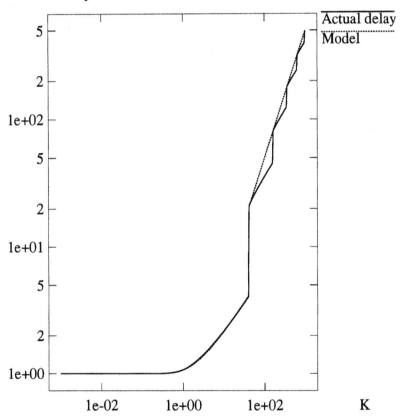

Figure 2.24 Composite delay model for all K values

2.10   SUMMARY

In this chapter, a number of recent techniques for the rapid simulation of MCM
interconnects were presented. These techniques achieve significant speedups
over conventional simulators by approximating the interconnects by lower-order
systems. They are very general, and can handle arbitrary tree structures of
lumped and distributed elements, crosstalk computation between coupled trees
and frequency-dependent effects. One of these techniques, Reciprocal Expansion (REX), supports incremental transfer function computation when the interconnection tree is modified, and is thus well-suited for guiding MCM physical
design. Delay modeling of MCM interconnects was also discussed, and a new
second-order delay model was presented to account for transmission-line effects
in the interconnect.

3

SYSTEM PARTITIONING AND CHIP PLACEMENT

3.1 INTRODUCTION

Partitioning and placement have been among the most popular research topics from the early days of CAD. Both problems are closely related - many placement algorithms are based on partitioning algorithms, and placement algorithms can also be used for effective system partitioning.

The partitioning problem arises in several important practical situations. The design of a complex system can be simplified considerably if the system can be decomposed into a number of smaller parts, which are relatively independent. This decomposition or partitioning procedure can be used during almost any stage of the design process. In its most general form, the partitioning problem can be stated as follows:

Given a graph $G(V, E)$,

Partition its vertex set into k subsets

Such that constraints on the sizes of the subsets are satisfied, and some objective is optimized.

The objective is typically to minimize or maximize the number (or weight) of edges "cut" by the partition, i.e., the edges connecting vertices in different subsets. The general problem is known to be NP-complete, although certain special cases are solvable in polynomial time.

A *min-cut* partition of a graph without restrictions on the sizes of the subsets can be found in polynomial time, using network flow techniques ([51]). However, in the presence of size constraints, the problem is intractable. One of the first effective heuristics for graph partitioning was proposed by Kernighan and Lin [52]. The algorithm was based on iteratively improving an initial partition, by exchanging subsets of vertices. The "K&L" algorithm was extended by Fiduccia and Mattheyses [53] to handle *hypergraphs*, which are a generalization of graphs in which a single *hyperedge* may connect several vertices. Hypergraphs are a better representation of circuits, which may contain multiterminal nets.

Real circuits usually contain "natural" partitions, which may divide the circuits into subsets of unequal size. The concept of *ratio-cut* partitioning was introduced in [54], which was able to find these natural partitions by softening the constraint on equal partition sizes. Cong, Hagen and Kahng demonstrated in [55] that partition quality can be further improved by considering the "dual intersection graph" of the netlist, instead of module adjacency graphs derived from the netlist hypergraph. A recent trend in partitioning is *timing-driven* partitioning, in which timing constraints are also included in the set of constraints to be satisfied.

In multichip system design, the system partitioning problem is encountered when a *full-custom* MCM design option is chosen, i.e., when the system designers choose to design a completely new chip set instead of using off-the-shelf parts. The problem then consists of partitioning the system design, which consists of a number of functional blocks, into chips, so as to meet constraints on the maximum area, power dissipation and I/O count on each chip, and also to meet timing constraints. The full custom approach can maximize system performance and density, at the expense of a longer design cycle time. Section 3.2 describes the MCM system partitioning problem and an algorithm to solve it.

Cell placement has also been a favorite research topic in VLSI physical design. Several placement approaches have been developed, including constructive placement, force-directed placement [56], simulated annealing [57, 58], simulated evolution [59], and min-cut placement [60]. The objectives of the classical VLSI cell placement problem have traditionally been minimization of chip area or wire length. In recent years, however, the emphasis has shifted to *performance-driven* placement, where the attempt is to minimize the contribution of interconnect delays to path delays. The reason for this trend is that improvements at the device and circuit levels have made the ratios of gate delays to interconnect delays smaller, thus placing the burden of performance optimization on the layout designers.

The importance of performance-driven layout is even greater in the design of MCMs. Here, due to the greater average length of the interchip connections, interconnect delays form a bottleneck impeding the system speed. In some ways, chip placement in MCMs is simpler than cell placement in VLSI. MCM chip placement involves fewer movable objects, and no wiring space needs to be allocated between the chips, since the wiring is done under them in the substrate. However, thermal constraints and the behavior of MCM interconnects require that new placement approaches be developed. For example, in VLSI, the delay of an interconnection net can be estimated by the semiperimeter of the bounding box enclosing all terminals of the net. However, in MCMs, the wire resistance cannot be neglected, and the delay becomes a function of the topology of the net, which is not easy to estimate during placement. Section 3.3 describes a performance-driven MCM chip placement algorithm which takes wire resistance into account to find placements with significantly improved performance.

3.2 MCM SYSTEM PARTITIONING

During the high-level system design phase, the system is specified by a set of interconnected functional blocks. The designers have rough estimates of the size, delay and power consumption of each functional block. The partitioning problem at this level consists of assigning these functional blocks to chips, subject to constraints on the size, power consumption and I/O pin requirement of each chip, and on signal delays between the functional blocks. For timing assurance, the partitioning step should ideally be linked to the low-level layout design, since signal delays between chips are a strong function of the chip placement.

The problem of performance-driven system partitioning for MCMs is discussed in [61, 62]. In [61], the problem formulation was introduced, and a heuristic algorithm was proposed to solve it, under two different delay models. In [62], the problem was generalized to handle arbitrary delay models, and formulated exactly as a Quadratic Boolean Programming problem. The following sections describe the problem formulation and solution strategies of [61] and [62].

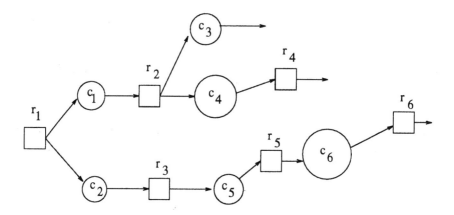

Figure 3.1 Synchronous system representation

3.2.1 Problem formulation

System representation

In the formulation of [61], the multichip system is assumed to be a synchronous digital system, consisting of a set R of register nodes and a set C of combinational blocks. The system is represented by a directed graph $G(V, E)$, where $V = R \cup C$ is the set of *nodes*, consisting of register nodes and combinational block nodes, and E is the set of *directed edges* representing signal flows between the system components (Fig. 3.1).

Each node $V_i \in V$ is associated with three attributes: area a_i, power consumption p_i and internal delay d_i. Each edge $e_{ij} \in E$ may represent either a single wire or a data bus with w_{ij} wires.

To simplify the problem formulation, it is assumed that each combinational block has exactly one input edge and one output edge. Further, each edge in E connects to a register node, i.e., every register-to-register path contains at most one combinational block.

Physical constraints

It is assumed that the chips will be assigned to a set S of slots arranged in a regular two-dimensional array on the MCM. The high-level partitioning problem is tied to the physical layout by including the slot assignment as a part of the partitioning process.

Each slot $s \in S$ can contain blocks with a maximum total area of A_s and a maximum total power dissipation of P_s. The total number of I/O pins in the slot can be at most IO_s.

Timing constraints

In the formulation of the timing constraints, it is assumed that the delay between blocks in two chip slots i and j is given by $D(i, j)$. Two different delay models are considered:

Constant Delay Model: The delay between any two chip slots is assumed to be a constant value.

Linear Delay Model: The delay between two chip slots is proportional to the Manhattan distance between them.

In practice, the actual delay of a multiterminal net is a more complicated function of the fanout of the signal and the actual topology of the net routing. For situations in which the wire resistance is significant, the delay of a point-to-point connection approaches being proportional to the *square* of the Manhattan distance between the signal source and sink. The timing formulation further assumes that the delay between functional blocks assigned to the same chip slot is negligible in comparison to the delay between slots. This assumption may not be realistic, especially in thin-film MCMs, where interchip wiring dimensions may be comparable to long on-chip wires. The larger size of off-chip drivers in comparison to on-chip drivers also contributes to reducing the gap between intrachip and interchip delays.

Mathematically, the formulation of the partitioning problem is expressed as follows: given a system graph $G(R \cup C, E)$ and a set S of slots, we seek a mapping $T : R \cup C \rightarrow S$, which minimizes the total number of interchip wire crossings:

$$W = \sum_{\forall i,j \ s.t. \ T(v_i) \neq T(v_j)} w_{ij}$$

subject to the following constraints:

Area Constraints:

$$\sum_{\forall i \; s.t. \; T(v_i)=s} a_i \leq A_s$$

Thermal Constraints:

$$\sum_{\forall i \; s.t. \; T(v_i)=s} p_i \leq P_s$$

I/O Constraints:

$$\sum_{\forall i,j \; s.t. \; T(v_i)=s, T(v_j)\neq s} w_{ij} \leq IO_s$$

Timing Constraints:

$$d_j + D(T(r_i), T(c_j)) + D(T(c_j), T(r_k)) \leq T_{cycle} \; \forall c_j \in C$$

where T_{cycle} is the system cycle time, and registers r_i and r_k are connected to the input and output of c_j, respectively.

3.2.2   Solution strategy

A heuristic algorithm for solving the partitioning problem under the *constant delay* model is described in [61]. The algorithm consists of the following basic steps:

1. **Super-node graph construction:** The nodes in the system graph are first merged into *super-nodes*. Each super-node consists of a set of nodes (functional blocks) which *must* be assigned to the same slot in order to satisfy timing constraints. Furthermore, any slot assignment which does not split a super-node is guaranteed to satisfy the timing constraints.

   Let the constant delay between slots be given by D. The super-node construction algorithm begins by setting each node $v_i \in V$ to be a super-node. It then looks at each combinational block $c_j \in C$ sequentially, and identifies three cases:

   If $T_{cycle} > d_j > T_{cycle} - D$: In this case, c_j as well as the registers r_i and r_k connected to its input and output must lie in the same slot to satisfy the timing constraints. Thus, the algorithm merges c_j, v_i and v_k into one super-node, where v_i and v_k are the super-nodes to which r_i and r_k belong.

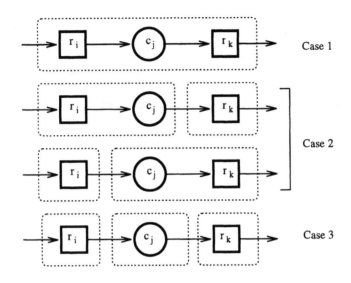

Figure 3.2   Super-node merging

If $T_{cycle} - D > d_j > T_{cycle} - 2 \times D$: Here, c_j must lie in the same slot as
either v_i or v_k. The algorithm merges it randomly with one of the two
super-nodes, and marks c_j as "semi-free," indicating that it may be
moved to the slot of the other super-node without timing violations.

If $d_j < T_{cycle} - 2 \times D$: In this case, no super-node merging occurs, since
the total delay is less than T_{cycle} even if C_j, r_i and r_k are all assigned
to different slots.

If $d_j > T_{cycle}$, no solution can satisfy the timing constraints, so the program
exits.

These cases are illustrated in Fig. 3.2. Note that the simplicity of the
super-node algorithm is made possible by the constant delay model and
the simplifying assumptions made on the system graph.

2. **K-way packing:** After constructing the super-node graph, the partitioning algorithm constructs an initial feasible partition (satisfying all the
constraints) by packing the super-nodes into the K slots. The packing
is done by first sorting the super-nodes in non-increasing order, and then
assigning them sequentially to slots. Each super-node is assigned to the
first slot which has enough capacity (area, power and I/O) remaining to
hold it.

3. **K-way slot assignment:** The initial feasible partition or assignment is improved by using the iterative exchange algorithm of Fiduccia and Mattheyses [53] to reduce the number of interchip wire crossings. The algorithm is modified to be able to handle the physical constraints. The "semi-free" blocks allow the algorithm greater flexibility to find better solutions. Note that the timing constraints are automatically satisfied, since super-nodes are not allowed to be split across slots.

The algorithm can also be extended to handle the linear delay model, using an approach similar to min-cut placement, as described in [63].

3.2.3 Quadratic Boolean Programming formulation

The performance-driven MCM system partitioning problem was generalized to handle arbitrary delay models, and reformulated as a Quadratic Boolean Programming (QBP) problem in [62].

An instance of the partitioning problem is described as follows:

System description: 1. A set J of N system components. Each component $j \in J$ has an associated size s_j, representing the silicon area required by j.

2. An $N \times N$ adjacency matrix A, where $a_{j_1 j_2}$ is the number of connections from component j_1 to j_2.

3. An $N \times N$ matrix D_C, where $D_C(j_1, j_2)$ is the maximum routing delay from component j_1 to j_2.

Module description: 1. A set I of M partitions or slots. Each slot i has a silicon capacity of c_i.

2. An $M \times M$ matrix B, where b_{i_1, i_2} is the cost of routing from slot i_1 to i_2.

3. An $M \times M$ matrix D, where $D(i_1, i_2)$ is the routing delay from slot i_1 to i_2.

4. An $M \times N$ matrix P, where p_{ij} is the cost of assigning component j to slot i.

The partitioning problem consists of finding an assignment $\mathcal{A} : J \to I$ which minimizes the total *cost*:

$$\sum_{\forall i,j \ s.t. \ \mathcal{A}(j)=i} p_{ij} + \sum_{\forall j_1, j_2} a_{j_1, j_2} b_{\mathcal{A}(j_1) \mathcal{A}(j_2)}$$

while satisfying the following constraints:

C1: Capacity constraints:

$$\sum_{\forall j \ s.t. \ \mathcal{A}(j)=i} s_j \leq c_i \quad \forall i \in I$$

C2: Timing constraints:

$$D\left(\mathcal{A}(j_1), \mathcal{A}(j_2)\right) \leq D_C\left(j_1, j_2\right) \quad \forall j_1, j_2 \in J$$

Notice that, as opposed to the earlier formulation, no assumptions are made about the structure of the circuit. Each component can have an arbitrary number of input and output edges, and a register-to-register path may contain an arbitrary number of combinational logic blocks. Furthermore, no relationship between the routing cost matrix B and the routing delay matrix D is assumed. Thus the delay model used can be as sophisticated as desired.

The new formulation can be recast *exactly* as a QBP problem, which is solved using a generalization of an existing powerful heuristic. However, there is a price to be paid for this increase in accuracy and flexibility: the delay constraints are not easy to compute from the system timing constraints under the general system model. System timing constraints are typically expressed in terms of upper (and sometimes lower) bounds on register-to-register delays. When register-to-register paths contain multiple combinational blocks, translating these specifications to upper bounds on individual net delays can be a difficult problem [64, 65, 66].

3.3 CHIP PLACEMENT

The chip placement and system partitioning problems are closely related. The partitioning approaches described in the previous section also generate a good relative placement for the chips in the final design. The placement approach described in this section can be used to simultaneously accomplish system partitioning, chip floorplanning and chip placement, when the input to the placement problem is specified in terms of the functional blocks of the system.

Performance-driven placement is very important in MCMs, since interconnect delays form a significant fraction of the system cycle time. In order to be effective, performance-driven design must be based on an accurate delay model. Until recently, reasonably accurate results could be obtained with a simplistic model in which the entire interconnection of a net was modeled as a single lumped capacitance, because driver impedances were large enough to swamp out the effects of resistance in the interconnect. Under the lumped capacitor model, the net delay is proportional to wire length; hence, delay minimization is equivalent to wire length minimization. Several performance-driven placement algorithms based on this delay model have been proposed [67, 68, 69, 70]. Timing constraints are translated to upper bounds on the lengths of nets, and these upper bounds are used to guide the placement algorithm.

However, for chip placement on MCMs, driver resistances are smaller (for greater speed), and interconnect lengths are an order of magnitude longer while feature sizes are comparable to on-chip wire widths (in MCM-D). Thus interconnect resistance cannot be ignored in performance-driven MCM design.

First-order RC delay models for resistive interconnects have been formulated by Elmore [49] and Rubinstein et al. [50] (Section 2.9.1 describes these delay models). Under the RC delay models, minimum wire length does not necessarily correspond to minimum net delay. The delay is a function of the net topology for multiterminal nets. This makes the placement problem more difficult, since net topologies are not known during the placement stage, whereas estimates of wire length can be made easily.

One way to get around this problem is to perform global routing simultaneously with the placement. Combined place-and-route approaches have been proposed previously [71, 72, 73]. The approach of [71] performs placement hierarchically, by recursively partitioning (*quadrisecting*) the layout area into four blocks. At each level of hierarchy, a global routing is performed, which is used to guide the partitioning and global routing at the next level.

In the following sections, we describe a quadrisection-based placement approach for MCMs, using a new objective function for the partitioning. Global routing is performed simultaneously with the placement, using a new type of tree called the *A-tree*, which has good delay properties when the interconnect is lossy.

Section 3.3.1 describes the delay model and objective function for the placement algorithm. Section 3.3.3 describes the placement algorithm, and Section 3.3.8 describes the global routing and terminal propagation techniques. Experimental results are presented in Section 3.3.9.

3.3.1 Delay model for lossy interconnect

When the total wire resistance of an interconnection tree becomes comparable to the resistance of the driver, the delay in the tree becomes a function of the topology of the tree, and not just the total wire length.

As described in Section 2.9.1, the distributed electrical parameters of an interconnection can be approximated closely by a tree T of lumped resistors and capacitors. The delay model of Eq. (2.82) can be used to estimate the interconnection delay:

$$t_d(T) = \sum_{k \in T} C_k R_{p(d,k)} \qquad (3.1)$$

where C_k is the value of the capacitance at node k. This model gives a single delay figure for all of the sink nodes, and is easier to use during physical design than the Elmore model.

It may be argued that inductive effects also become significant in MCMs, and a second-order delay model may be more appropriate. However, inductive effects are too intimately linked to the actual net topologies and electrical parameters to be accurately accounted for during the placement stage. Thus, we restrict ourselves here to a first-order delay model.

Consider an RC tree on a regular global routing grid, as shown in Fig. 3.3. Each node k in the tree T is associated with a unit length of wire, which has a resistance of R_0 and capacitance of C_0. As described in [74], the delay of Eq. (2.82) can be split into four terms and rewritten as

$$t_d(T) = t_1(T) + t_2(T) + t_3(T) + t_4(T) \qquad (3.2)$$

where

$$t_1(T) = R_d \sum_{k \in T} C_0 = R_d C_0 |T|$$

$$t_2(T) = R_0 \sum_{s \in sink(T)} l_{p(d,s)} C_s$$

$$t_3(T) \quad = \quad R_0\, C_0 \sum_{k \in T} l_{p(d,k)}$$

$$t_4(T) \quad = \quad R_d \sum_{k \in sink(T)} C_k \ = \ \text{constant}$$

where R_0 and C_0 are the resistance and capacitance per unit length of the wire, R_d is the driver resistance, C_s is the gate capacitance at sink node s, and $l_{p(d,k)}$ is the length of the path from the driver to node k.

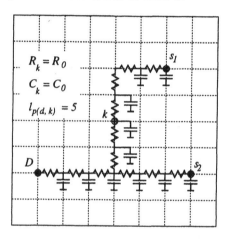

Figure 3.3   RC tree on a global routing grid

The first term is proportional to the total number of nodes in the tree, or equivalently, the total wire length of the tree. The second term is proportional to the sum of driver-to-sink path distances. The third term is related quadratically to the path lengths in the tree. Since the fourth term is a constant, it can be ignored for optimization purposes. The *cost* (delay) of a net can be viewed as being constituted of two contributions: one due to the *segments* or units of wire which form the net, and the other due to the sink capacitances. The segment cost contribution can be expressed as follows:

$$c_1(T) = t_1(T) + t_3(T) \quad = \quad \sum_{k \in T} \left[R_d\, C_0 + R_0\, C_0\, l_{p(d,k)} \right] \qquad (3.3)$$

$$= \quad \sum_{k \in T} \left[LWC(k) + QWC(k) \right] \qquad (3.4)$$

By definition, segment k is the wire segment joining node k to its parent in the interconnection tree. Thus, each segment k contributes a *linear wire cost*

(LWC) equal to $R_d C_0$, plus a *quadratic wire cost (QWC)*, which is given by $R_0 C_0$ times the distance $l_{p(d,k)}$ between node k and the root.

The second cost contribution, due to the sink capacitances, is given by the sum of all driver-to-sink path distances, multiplied by $R_0 C_s$, where C_s is the load capacitance at each sink:

$$c_2(T) = t_2(T) = R_0 \sum_{s \in sink(T)} l_{p(d,s)} C_s \qquad (3.5)$$

3.3.2 Path delay vs. net delay

In typical digital systems, which consist of sequential synchronous logic circuits, timing constraints are expressed in terms of *latch-to-latch delays* [75], i.e., the total delay of a signal path from one clocked storage element to the next should lie within a specified range (Fig. 3.4). Since a latch-to-latch path may consist of several gates and nets, minimizing the average net delay does not guarantee that all timing constraints will be satisfied. In recognition of this fact, most recent work on performance-driven VLSI physical design has been focused on meeting path delay constraints [68, 69, 70, 76].

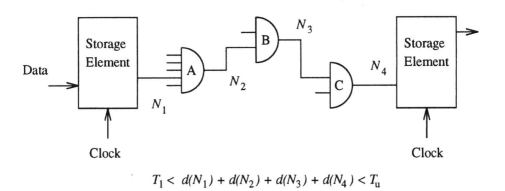

$$T_1 < d(N_1) + d(N_2) + d(N_3) + d(N_4) < T_u$$

Figure 3.4 Timing constraints in synchronous circuits

At the system level, however, assuming a well-partitioned system, it is very unlikely that there will be more than one interchip connection between a pair of latches. Thus, in MCM physical design, latch-to-latch delays can be minimized by minimizing interchip net delays. The overall system speed can be maximized by minimizing the *maximum* net delay.

The placement algorithm presented in this chapter attempts to minimize the total cost $c_1(T) + c_2(T)$ of all the nets, or equivalently, to minimize the average interchip net delay. However, the quadrisection-based placement algorithm allows considerable flexibility in choosing the cost function. It is possible to use different weights, or even different cost functions, for different nets. Thus, the maximum net delay can be minimized by updating the weights of nets at the end of each level of partitioning, assigning higher weights to nets with longer estimated delays. Path delay-constrained placement can also be implemented by computing delay bounds for all nets initially, and iteratively updating the weights of nets after each level of partitioning, assigning a higher weight to nets with a smaller estimated slack as in [68]. It is also possible to specify different electrical parameters for different nets, or to use the delay objective for a small subset of the nets, and a more conventional objective such as min-cut or wire length for noncritical nets.

3.3.3   Resistance-driven placement algorithm

Min-cut quadrisection has been demonstrated to be an effective technique for two-dimensional placement problems [77]. In this approach, the set of cells to be placed is partitioned into four subsets, each of which is assigned to one quadrant of the layout area. Each subset is then recursively partitioned into four smaller pieces, until the position of each cell has been determined. It is possible to combine the hierarchical placement step with the global routing step, by hierarchically constructing global routing trees, as described in [71].

Following the terminology of [77], we denote the set of nets by N and the set of cells (or chips, in the case of MCM placement) by C. The set of cells in a net n is denoted by C_n, and the set of nets connected to a cell c is denoted by N_c. A cell containing the driver pin of net n is called the *driver* of net n and denoted by d_n. If k is the average number of terminals in a net and p is the average number of nets connected to a cell, the size of the input is given by

$$M = k|N| = p|C|$$

At each level of quadrisection, each set of cells to be partitioned is referred to as a *block*. After one or more stages of quadrisection, the cells connected to a net may be distributed among different blocks. The set of blocks containing one or more cells in C_n is denoted by B_n.

During quadrisection, each block is partitioned into four subblocks, denoted by W, X, Y and Z (Fig. 3.5). The *cell distribution* of a net in a particular block $B \in B_n$ is a vector

$$\alpha(B, n) = < \alpha_W(B, n), \alpha_X(B, n), \alpha_Y(B, n), \alpha_Z(B, n) >$$

where $\alpha_K(B, n)$ is the number of cells connected to net n in subblock K of block B, $K \in \{W, X, Y, Z\}$.

Block B

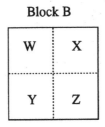

Figure 3.5   Quadrisectioning of a block B

We define another vector $\Pi(B, n)$ associated with each block $B \in B_n$, which contains the following information:

1. K_d, the subblock which is closest to d_n. If the block B contains d_n, then K_d is identical to the driver subblock.

2. D, the distance between K_d and d_n.

3.3.4   Cost function

In the conventional min-cut quadrisection procedure, the cost contribution of a net during the partitioning of a block B is a function of $\alpha(B, n)$. The cost depends only on whether the net has been "cut" by the vertical and horizontal cut lines.

In our cost formulation, the cost contribution of a net n in a block B is a function of both $\alpha(B, n)$ and $\Pi(B, n)$. Since D and K_d in $\Pi(B, n)$ depend on the position of the driver d_n, which may not be in B, the cost depends on the partitioning within the block containing d_n.

Consider a block as shown in Fig. 3.6. The α values for a net n are indicated within the respective subblocks. The global routing tree constructed after the previous level of partitioning is also shown. In the figure, the driver of the net is assumed to be outside the block, at a distance D from the subblock K_d. All cells within a subblock are assumed to be located at the center of the subblock. The distance between subblocks is computed as the center-to-center Manhattan distance. The physical length corresponding to one grid unit is updated after each level of quadrisection, in order to ensure correct delay computation.

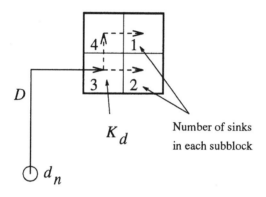

Figure 3.6   Cost computation for a net

The computation of the three cost components of a net n, described in Section 3.3.1, is accomplished as follows:

Linear Wire Cost (LWC): This term is proportional to the total wire length required for the intrablock connections. Each segment of wire used to complete the connections contributes a cost of $R_d C_0$. This computation depends on the $\alpha(B, n)$ values and K_d. In the example, $LWC = 3R_d C_0$, since three wires of unit length (shown in dotted lines) are required for the intrablock connections.

Driver-Sink Distances (DSD): This term computes the sum of the distances from the sink nodes to the driver node. This computation depends on the $\alpha(B, n)$ values, D and K_d. In the example, there are three sinks inside the block located at a distance of D from the driver node, $(2 + 4)$ sinks located at a distance $D+1$, and one sink at a distance of $D+2$, giving a total cost of $DSD = R_0 C_s \left(3D + 2(D + 1) + 4(D + 1) + 1(D + 2)\right)$.

Quadratic Wire Cost (QWC): Each segment of intrablock wire contributes a quantity to this term, equal to $R_0 C_0$ times the distance of the end point of the segment from the driver subblock. This computation also depends on the $\alpha(B, n)$ values, D and K_d. In the example, there are two wire segments whose end points are located at a distance of $D+1$ from the driver, and one segment (the upper horizontal segment) whose end point is $(D+2)$ units away from the driver. Thus, $QWC = R_0 C_0 (2(D+1) + (D+2))$.

The total cost contribution of a net n is the sum of LWC, DSD and QWC:

$$w(\alpha(B, n), \Pi(B, n)) = LWC + DSD + QWC$$

3.3.5 Partitioning algorithm

An effective linear-time algorithm for network bisection was proposed by Fiduccia and Mattheyses [53], and extended for quadrisection by Suaris and Kedem [77]. The basic idea of the algorithm is as follows: for each cell c in subblock $U \in \{W, X, Y, Z\}$ of the block to be partitioned, a *gain* value $G^{UV}(c)$ is computed for each $V \in \{W, X, Y, Z\}, V \neq U$. This gain equals the change in total cost when cell c is moved from subblock U to subblock V. The gain values are stored in 12 *gain tables*, specialized data structures which allow efficient updates of the gain values. Once the gain tables have been initialized, the algorithm proceeds to move cells one at a time. At each step, the cell with maximum gain $G^{UV}(c)$ is taken out of subblock U and locked in subblock V. After each move, the gains of other free (unmoved) cells are updated. The procedure continues until no free cells are available. The algorithm is effective because it allows "uphill" moves also, allowing it to escape from local minima.

3.3.6 Gain computation

The computation of the values $G^{UV}(c)$ in our placement algorithm is slightly different from the original quadrisection algorithm. This is due to the fact that in our algorithm, a move within the block being partitioned may have repercussions outside the block. For instance, when a chip containing the driver of a net n is moved, the D and K_d values may change in all blocks in B_n.

Let B be the block currently being partitioned, and let c be some cell in subblock U of B. Each net n in N_c contributes to $G^{UV}(c)$. If $c \neq d_n$, then the contribution is given by

$$w(\alpha^{UV}(B,n), \Pi(B,n)) - w(\alpha(B,n), \Pi(B,n)) \tag{3.6}$$

where $\alpha^{UV}(B,n)$ is the new cell distribution vector after c is moved from U to V. Note that $\Pi(B,n)$ does not change.

If $c = d_n$, the cost computation will involve all blocks in B_n, in general. The gain of moving d_n from U to V will be given by

$$\sum_{A \in B_n, A \neq B} [w(\alpha(A,n), \Pi^{UV}(A,n)) - w(\alpha(A,n), \Pi(A,n))]$$
$$+ w(\alpha^{UV}(B,n), \Pi^{UV}(B,n)) - w(\alpha(B,n), \Pi(B,n)) \tag{3.7}$$

Here, $\Pi^{UV}(A,n)$ contains the updated values of D and K_d for block A after d_n moves from U to V in block B. Note that $\alpha(A,n)$ does not change for $A \neq B$.

3.3.7   Data structures and complexity analysis

The linear time complexity of the original quadrisection algorithm is made possible by the special structure of the gain tables, which allows $O(1)$ time access of the cell with maximum gain value, and $O(1)$ time update of the gain value of any cell. The data structure made use of the fact that the gains were integers, and were bounded between two limits. Unfortunately, the gain values under our cost formulation are nonintegers, so the same data structures cannot be used. At present, the best we can do is to store the gain values in a heap or a height-balanced tree. This gives us constant-time access to the cell with maximum gain, and $O(\log |C|)$ time for updating the gain value of any cell, or deleting the entry corresponding to a cell after it is locked in place.

The quadrisection algorithm we use is very similar to that of [77]. We begin with an initial partition satisfying the size balance constraints and then attempt to improve it by repeated applications of a procedure *pass*, as illustrated in Fig. 3.7.

The procedure *pass* begins with an initialization of the 12 gain tables, one for each move direction. Consider the time taken for computing the contribution of a single net n to the gain values of cells in C_n. For the driver, this can take $O(|C_n|)$ time in the worst case, since computing the driver gain involves all blocks in B_n. For the other cells, the gain can be computed in $O(1)$ time, so the contribution of net n to the gains of all cells in C_n can be computed in $O(|C_n|)$ time. Therefore, the total time taken for computing the gains is $O(|N|k) =$

```
quadrisection(){
      P := initial partition;
      do{
            P := pass(P);
      }while(cost improved);
}


pass(P){
      initialize α values;
      initialize gain tables;
      optimal = W(P);
      while(there is an allowed move){
            c, U, V := cell and direction with highest gain;
            gain := G^{UV}(c);
            remove c from gain tables;
            lock c in V;
            for each n ∈ N_c{
                  for each c' in C_n: update gain(c');
                  for each B ∈ B_n: update α(B,n), Π(B,n);
            }
            W(P) := W(P) - gain;
            if(W(P) < optimal){
                  optimal = W(P);
                  delete stack;
            }
            else add c, U, V to stack;
      }
      undo moves in stack;
}
```

Figure 3.7   The quadrisection algorithm

$O(M)$. Inserting the cell gain values in the gain tables takes $O(|C| \log |C|)$ time, if the tables are implemented as heaps or height-balanced trees.

After the gain tables are initialized, an iterative improvement loop is repeated until no movable cells remain. In each iteration, the cell c and move direction

UV with highest gain are selected, cell c is locked in subblock V, and c is removed from the gain tables. This procedure takes $O(\log|C|)$ time. After c is moved, the gains of all cells connected to c are updated, and the α and Π values of the nets in N_c are updated.

When a cell c is moved, $|N_c|$ nets are affected, and for each affected net n, $O(|C_n|)$ chips' gains have to be updated. Each update takes $O(\log|C|)$ time. Therefore, using a very simple analysis, the complexity of *pass* is $O(kp|C|\log|C|)$, which can be written as $O(kM \log(M/p))$. In typical circuits, k (the average number of terminals in a net) is a small number which does not vary much with problem size. The procedure *pass* is typically executed a small number of times before it converges, so the overall complexity of the quadrisection procedure is $O(kM \log(M/p))$.

3.3.8   Terminal propagation

While partitioning a block at a particular level of the hierarchy, it is important to consider the locations of cells outside the block as well. This is accomplished using a technique known as *terminal propagation* [78]. Also, to reduce the dependency on the sequence in which the blocks are chosen for partitioning, it is helpful to iterate the partitioning sequence a few times, until no further reduction in cost is obtained. The combination of these two enhancements leads to significant improvements in the placement quality [77].

We denote the set of terminals of a net lying outside the block being partitioned as *exterior terminals*. In terminal propagation, exterior terminals of a net are taken into account by introducing "dummy terminals" on the perimeter of the block. For example, in Fig. 3.8, the exterior terminals a, b and c of the net are propagated to dummy terminals a', b' and c' on the boundary of block B, in order to influence the partitioning of B.

One important point to consider during terminal propagation is which subset of the exterior terminals to propagate to a block. Clearly, propagating *all* the exterior terminals of a multiterminal net may not be effective. Terminal propagation is most effective when it is guided by simultaneous global routing, since only those exterior terminals which are directly connected to terminals within the block are propagated. For example, only terminal a is propagated to block B in Fig. 3.9, resulting in improved partitioning of B.

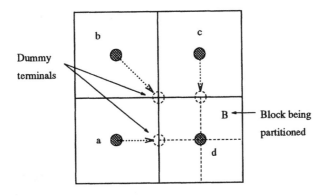

Figure 3.8 Terminal propagation

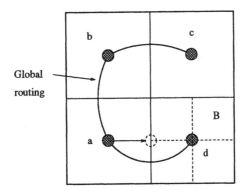

Figure 3.9 Terminal propagation guided by global routing

Terminal propagation in our resistance-driven placement algorithm is enhanced by simultaneous global routing. The global routes are constructed hierarchically: after all blocks at the current level of the partitioning hierarchy have been processed, the global routing generated at the previous level is updated, as described below. In keeping with the performance-driven design philosophy, special trees called *A-trees* are used for the global routing. These trees have been shown to reduce delays significantly in comparison with minimum wire length Steiner trees [74]. The fundamental property of an A-tree is that every vertex in the tree is connected to the root vertex by a shortest-length path. A-trees are described in greater detail in Chapter 5.

We denote by $T_i(n)$ the tree constructed for net n after the ith level of partitioning. Thus $T_0(n)$ is a trivial tree, consisting of a single node. At any level,

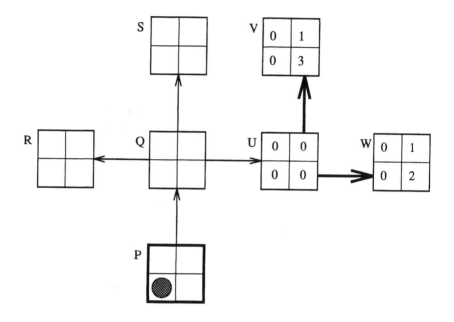

Figure 3.10 Global routing tree for a net

each node in $T_i(n)$ corresponds to a block B at level i. After partitioning at level $i + 1$, each node in $T_i(n)$ may have to be split into as many as four new nodes, to obtain $T_{i+1}(n)$. We will now describe how this is accomplished.

Figure 3.10 shows a tree with seven nodes (blocks). Node P is the root or driver node. The position of d_n after the partitioning is shown by the circle. The numbers in blocks U, V and W show the α values within these blocks. Note that, due to the A-tree restriction, the following statements must be true:

1. A node which is in the same row or column as the root node can have at most three children (node Q in the figure).

2. A node which is not in the same row or column as the driver can have at most two children. Furthermore, the tree edges connecting the node to its children must be directed *away* from the root (e.g., node U in the figure).

The root node can have at most four children. The nodes of $T_i(n)$ are processed in a reverse depth-first traversal. Thus, while processing a particular node x, the locations of the connections to the children of x have already been fixed. For example, for node U in the figure, it has been decided that the connection to V will come out of the *right* upper subblock, instead of the left, and the connection to node W will come out of the *lower* right subblock, instead of the upper.

Consider a block B, corresponding to a node x in T_i, which has been partitioned into four subblocks. The node x will be deleted, and replaced by up to four new nodes. The number of new nodes which will be created depends on the number of subblocks in B which are *occupied*. A subblock is defined to be "occupied" if at least one of the following conditions holds:

■   The subblock has a positive α value, i.e., it contains at least one cell of net n, or

■   A connection to a child node comes out of the subblock.

Thus, the right upper and right lower subblocks of block U are defined to be occupied, even though they do not contain any cells of n.

Each occupied subblock will be represented by one new node in T_{i+1}. Of these new nodes, the node closest to the driver will become the new child of the parent of x. This node is denoted the *eldest node*. If there are two nodes at equal distance from the driver, as in Fig. 3.11, an extra node has to be added as the eldest node.

If the partitioning and application of the above rules result in the generation of more than one new node, they have to be attached to the tree as descendants of the eldest node. Nodes in the same subrow or subcolumn as the eldest node are connected as its children, and called *intermediate nodes*. The node diagonally opposite the eldest node is called the *youngest node*. If both of the intermediate nodes are absent, the youngest node is connected directly as a child of the eldest node. Otherwise, the youngest node is connected as a child of the intermediate node with a higher α value (Fig. 3.12). If both intermediate nodes have the same α value, one of them is chosen arbitrarily.

It can be verified that the above rules for updating the global routing preserve the A-tree nature of the global routes.

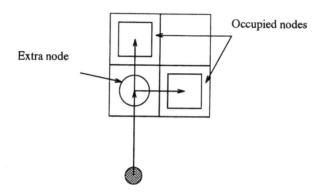

Figure 3.11   Addition of an extra "eldest" node

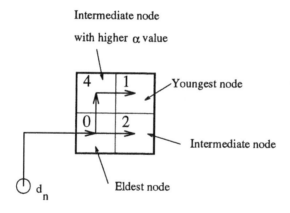

Figure 3.12   Intracell global routing

3.3.9   Experimental results

The new placement algorithm has been implemented in C on a Sparcstation 1 running Unix. To verify that the approach does indeed produce placements with lower delays, several random netlists were generated, with up to 1024 movable cells. To simplify the initial testing, the number of cells was constrained to be a power of 4, and the cell sizes were assumed to be equal. These assumptions, however, are not necessary for the placement algorithm. The netlists had several multiterminal nets with three to seven terminals. The electrical parameters used corresponded to a thin-film MCM situation, with $R_d = 200\Omega, R_0 = 20\Omega, C_0 = 0.133\ pF$, and $C_s = 1.0\ pF$, where R_0 and C_0 values are per unit chip pitch. The net delays were computed by hierarchically constructing A-*trees* simultaneously with the placement as described in Section 3.3.8, and applying the delay model of Eq. (2.82) to these trees. The global routes can also be constructed upon completion of placement, using the algorithms described in Chapter 5.

For comparison, we also implemented three other cost functions in addition to the new objective. The first was a wire length cost function, which consisted of only the LWC term of the delay cost. The second was the conventional min-cut cost function, defined as follows: the cost of a net N in a partition is 0 if all terminals of N lie on the same side of the cut line; the cost is 1 otherwise. The min-cut cost function was guided by terminal propagation.

The min-cut objective has been recognized to have certain limitations [79]. For example, consider the situation in Fig. 3.13. The cost of the net under the min-cut objective is 1. Since moves M_1 and M_2 do not affect the cost of the partition, they are both equally favored during the partitioning process. However, it is evident that move M_1 is more desirable than M_2, since M_1 brings the partition closer to a zero-cost partition, whereas M_2 moves it farther away.

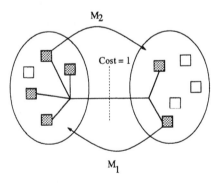

Figure 3.13 Limitation of min-cut objective

A new cost function which recognizes this limitation is proposed in [79], for a row-based placement problem. We have implemented a simpler variation of this objective, which we call the "cut-ratio" objective, defined as follows: we define h_{min} to be the smaller of the number of cells above the horizontal cut line and the number of cells below the cut line, i.e.,

$$h_{\min} = \min\{\alpha_W + \alpha_X, \alpha_Y + \alpha_Z\}$$

Similarly, the quantity $v_{\min}$ is defined as

$$v_{\min} = \min\{\alpha_W + \alpha_Y, \alpha_X + \alpha_Z\}$$

The cut-ratio cost is defined as

$$C_{cr}(\alpha) = \frac{2 \times (h_{\min} + v_{\min})}{\alpha_W + \alpha_X + \alpha_Y + \alpha_Z} \qquad (3.8)$$

Note that the cut-ratio objective is not related to the "ratio-cut" partitioning approach [54]. In ratio-cut partitioning, the objective is to minimize the total *number* of nets cut, divided by the product of the sizes of the partitions. In our approach, the partition sizes are constrained to be nearly equal. The cut-ratio objective was also guided by terminal propagation.

The experimental results are presented in Table 3.1. The first two columns give the number of cells and nets in each example. The next four columns give the maximum net delays for placements using the four cost functions. The delays were computed using Eq. (2.82). The maximum net delay figures are significant, since the system speed is limited by the slowest net. The average and maximum net delays were always the least when the new delay objective function was used, except for the fourth example, in which the wire length minimization produced a smaller maximum net delay. The improvement becomes even more significant as the problem size increases. For the larger examples, the maximum net delay was *as much as 50% more* using the min-cut and ratio-cut cost functions. It should be noted that the terminal propagation for *all* cost functions was guided by the simultaneous hierarchical A-tree global routing, so the large delays for the wire length, min-cut and ratio-cut approaches were due entirely to the fact that these cost functions ignored the interconnect resistance.

In some situations, the area penalty caused by focusing on delay minimization, or the extra computational effort required, may be unacceptable. To assess the penalty caused by using the delay objective, we also compared the wire lengths of the global routing trees in the placements and the execution times using the different cost functions. These comparisons are presented in Table 3.2. As the figures show, the wire lengths are generally the least when the wire-length cost function is used, as expected. What is perhaps surprising is that the wire lengths obtained using the delay cost function are actually *smaller* than those obtained using the min-cut and ratio-cut cost functions. Therefore, the area penalty using our approach will be small.

The last four columns in Table 3.2 give the run times for the algorithm using the different cost functions. The run times for the other cost functions were approximately half the run times for the delay cost function (the min-cut algorithm with the original data structures would probably be even faster). However, the difference in run times is more than offset by the significant delay reductions, especially for high-performance system design.

Table 3.1　Delay results on eight examples

Cells	Nets	Maximum Net Delay (ns)			
		LDQ^1	LWC^2	Min-Cut	Cut-Ratio
64	60	**2.507**	3.011	3.136	2.519
64	80	**2.753**	2.990	3.166	2.805
64	100	**2.794**	3.195	3.246	2.857
256	200	4.332	**4.273**	5.786	4.968
256	250	**4.404**	4.917	6.259	4.512
256	300	**4.272**	6.723	5.863	5.405
1024	1000	**9.108**	11.36	14.10	12.98
1024	1200	**9.264**	13.06	16.05	15.14

[1] Delay minimization using $LWC + DSD + QWC$
[2] Wire length minimization

Table 3.2　Global routing wire lengths and run times

Total Wire Length				Run Time (s)			
LDQ	LWC	Min-Cut	Cut-Ratio	LDQ	LWC	Min-Cut	Cut-Ratio
422	408	498	470	10.4	7.3	3.9	7.9
585	608	674	628	17.2	8.9	7.7	14.3
801	781	836	852	21.9	14.9	10.0	11.8
2430	2449	2986	2651	53.6	33.6	22.4	29.2
3240	3194	3707	3646	80.9	43.8	29.6	35.7
4245	4067	4656	4341	87.0	50.3	38.1	52.4
27388	25504	30304	27709	520.5	303.6	189.7	267.3
35601	33579	38546	35910	616.4	318.1	268.7	286.2

3.3.10   Variance minimization

If the weights of all of the nets are equal, the placement algorithm minimizes
the *average* net delay. However, as described in Section 3.3.2, the system performance is maximized when the *maximum* net delay is minimized. The results
presented in Table 3.1 demonstrate significant reductions in maximum net delay even without any net weighting. However, these results can be improved
even further, by updating the weights of nets after each level of partitioning.
By increasing the weights of nets with greater estimated delays, the placement
algorithm can be forced to give more importance to reducing the maximum
delay.

The placement algorithm was tested on the examples of Table 3.1, using the
following update scheme for the net weights: the initial weight for each net was
set to 1. After each level of partitioning and global routing, the weight of each
net was set equal to its estimated delay. Thus, the algorithm was effectively
minimizing the *mean squared delay* of all the nets, i.e., the objective of the
algorithm can be expressed as

$$\min \ \sigma^2 + \mu^2$$

where σ^2 is the variance of the net delays and μ is the mean. This scheme
increases the effectiveness of the placement algorithm significantly. Figure 3.14
shows the maximum net delays for the eight placement examples, with and
without the net weighting scheme. The delay using mincut placement is also
shown for reference.

3.4   SUMMARY

The related problems of system partitioning and chip placement for MCMs were
discussed in this chapter. Recent approaches to solve the MCM system partitioning problem, subject to a variety of practical constraints, were presented.
A new "resistance-driven" placement algorithm for MCM net delay minimization was described. The algorithm is able to achieve significant performance
improvements over conventional min-cut and wire length reduction-based placement techniques, by taking wire resistance into account while placing the chips.
The placement algorithm can also be used for effective system partitioning.

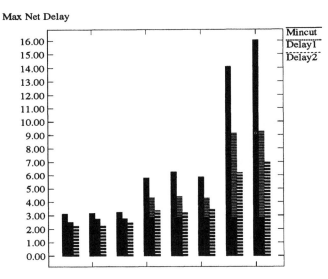

Delay1: Without net weighting
Delay2: Net weight = estimated delay

Figure 3.14 Maximum delay reduction using variance minimization

MULTILAYER MCM ROUTING

4.1 THE MULTILAYER ROUTING PROBLEM IN MCMS

The problem of routing the interchip connections in an MCM substrate differs from conventional on-chip routing in some fundamental ways. The very nature of the routing environment is different: in most VLSI design styles, such as the standard cell, sea-of-gates and macro-cell, the area available for routing intercell connections is divided into *channels*, and *switchboxes* as illustrated in Fig. 4.1. The routing process is thus naturally decomposed into two stages: *global routing*, in which routes for all nets are assigned to appropriate sequences of channels; and *detailed routing*, in which the wires in each channel and switch-box are ordered and assigned to specific layers such that design rule constraints are satisfied. Typically, there are only two or three layers available for signal wiring. These may differ significantly in their electrical properties, such as the resistance per square of the interconnect and the capacitance per unit area.

In MCMs, on the other hand, we have a *general area* routing problem [21]. The chips placed on the top layer do not interfere with the substrate wiring, thus the routing area does not get decomposed into channels (Fig 4.2). The entire rectangular region is available for signal wiring. However, there may be obstacles introduced by thermal vias used for conducting heat away from the chips through the substrate, or by power and ground connections. The number of layers may be very large, especially in ceramic (MCM-C) modules, and to a lesser extent in laminated (MCM-L) modules. For example, the ceramic module (S/390 alumina TCM) used in IBM's Enterprise System/9000 proces-

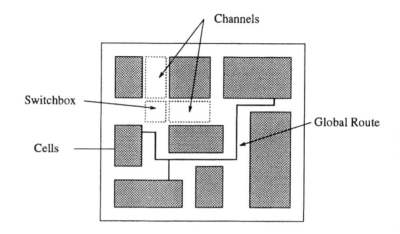

Figure 4.1   VLSI routing environment

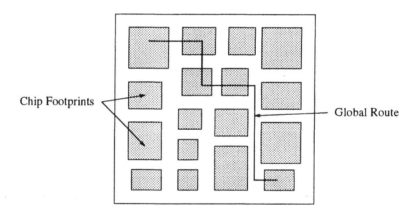

Figure 4.2   MCM routing environment

sor family contains 63 metallized ceramic wiring layers [4], and the VP-2000 supercomputer from Fujitsu uses a ceramic module with over 50 layers [80].

The multilayer substrates in MCM-C and MCM-L also allow the use of *blind*, *buried* and *stacked* via structures, while MCM-D substrates allow *staggered* vias (Fig. 4.3). Such structures are not generally available on conventional PCBs, which support only *through-hole* vias (Fig. 4.3.a). A through-hole via can be used by only one net, whereas a segmented via can be used by different nets between different, nonoverlapping sets of layers. These via structures can

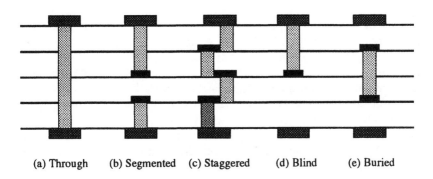

(a) Through   (b) Segmented   (c) Staggered   (d) Blind   (e) Buried

Figure 4.3 Via structures in MCMs

enhance the routing efficiency, and specialized routing algorithms are required to take advantage of them.

Another important difference between VLSI and MCM routing is the behavior of the interconnects. As mentioned in Chapter 2, transmission line effects become noticeable in interchip routing. Since performance is a critical issue in MCMs, routing algorithms must take into account such issues as reflections, crosstalk. In contrast, the higher driver impedances and shorter lengths of on-chip connections make the consideration of such effects less important in VLSI routing.

4.2 APPROACHES TO MCM ROUTING

Due to the large number of complex, high-pinout chips in many MCM designs, the amount of interchip wiring can be very large. For example, IBM's S/390 alumina TCM [4] contains 400 m of wiring in the substrate. In order to complete the routing of these connections, while simultaneously satisfying manufacturability, reliability and crosstalk constraints, a large number of layers may be required. This gives rise to a routing problem which is almost three-dimensional in nature. (The problem is "almost" three-dimensional because net terminals can only lie on the top or bottom layers. However, future advances in packaging, such as stacking of several chip layers, may lead to true three-dimensional routing problems.)

There are several possible approaches to this three-dimensional routing problem. One approach is direct maze-routing on a 3-D grid. Despite the conceptual simplicity of this approach, it has several drawbacks. The first problem is the large memory requirement. To illustrate the severity of this problem, consider an industrial MCM benchmark circuit released by MCC: this example has a routing area of 174 mm × 174 mm, and a routing pitch of 75 microns, and requires about seven layers to complete the routing [21]. This leads to a total grid size of 28×10^6! Clearly, storing and searching in such a large grid will be extremely inefficient. Even if it were possible, it is well-known that even for two-dimensional maze routing, the completion rate is strongly dependent on the ordering of the nets, and no good algorithms are known for finding an optimal or even a good ordering in general. Furthermore, in maze routing, it is extremely difficult to impose additional constraints, such as performance. Some nets may be forced to take long paths, with many bends, resulting in poor timing performance.

This chapter discusses some recent multilayer routing algorithms developed specifically for MCM designs. Section 4.3 describes the *SLICE* router [21], which decomposes the 2-D grid into narrow "slices", and is thus able to handle large and dense designs. The "V4R" router [81] which guarantees that every net is routed using at most four vias, is described in Section 4.4. "Rubber-band" routing [82] is described in Section 4.5.

4.3   THE SLICE ROUTER

An efficient multilayer general area routing algorithm for MCMs was presented in [21]. The router is called SLICE, as it operates on narrow slices of the routing area. The router proceeds in a layer-by-layer order: it first tries to complete the routing of as many nets as possible on the top layer. The nets which could not be routed, or which could only be routed partially, are then propagated to the next layer, and the process is repeated. The output of the router is a set of wire segments and vias connecting the terminals of each net. The completion rate is always 100%, since the number of layers is not prespecified: the router uses as many layers as necessary to complete the routing. In practice, manufacturing and cost constraints may specify a limit on the number of layers, and unrouted nets will have to be manually completed.

The SLICE router is able to handle a number of practical MCM physical features, such as stacked and segmented vias (Fig. 4.3) and routing blockages caused by power and ground connections and thermal vias. The formulation is intended only for two-terminal nets. This restriction is claimed to have only

Algorithm SLICE

    N = list of nets;

    $l = 1$;   /* Start at top layer */

    while $(N \neq \emptyset)$ **do**   /* while some unrouted nets remain */

        Compute planar routing on layer l;

        Redistribute uncompleted terminals on layer l;

        Perform restricted maze routing on layer l;

        Remove jogs;

        Propagate uncompleted net terminals to layer $l + 1$;

        Rotate the routing grid by 90°;

        $l = l + 1$;

    end

end

Figure 4.4 Overview of SLICE [21]

a small effect on the solution quality, since most MCM nets are two-terminal nets, and k-terminal nets can be decomposed into $k-1$ two-terminal nets based on a minimum spanning tree. The algorithm also allows routes corresponding to subnets of the same net to touch and merge into a Steiner tree. The basic approach assumes a Manhattan routing geometry, although it can be extended to handle 45° routing. SLICE runs six times faster than 3-D maze routing, and uses much less memory, since it does not operate on the entire routing grid. The routing solution is also superior to maze routing in terms of the total number of vias.

4.3.1 Overview

The SLICE algorithm consists of the following basic steps, which are repeated on every layer (Fig. 4.4):

1. **Planar routing:** This is the most important step in the process. SLICE attempts to complete the routing for as many nets as possible in the current layer, so as to minimize the number of layers as well as the number of vias. The nets which cannot be completed are routed partially so that they can be completed more easily in the next layer, with shorter wires.

The routing grid on the current layer is divided into vertical channels by "columns," where a column is a vertical grid line containing at least one net terminal. A topological routing is computed in each channel, starting from the leftmost channel. Then a physical routing is generated corresponding to the topological routing. The channels are processed one at a time, from left to right.

2. **Pin redistribution:** After the planar routing step is completed for all the channels, some of the nets are only partially routed. The terminals of the partial routes have to be propagated to the next layer, so that the router can attempt to complete their routing later. The propagated terminals tend to be clustered, since the same obstacles which prevented the completion of a particular net would also have obstructed several other nets. To avoid congestion and blockages in the next layer, the clustered terminals are redistributed by moving them to nearby columns with fewer terminals, so that the number of terminals in each column is made roughly equal and the increase in wire length due to the redistribution is minimized.

3. **Restricted maze routing:** The division of the 2-dimensional grid into columns biases the routing procedure on the current layer in the horizontal direction. In order to increase the chances of completion of nets which are nearly vertical, a maze routing step is performed after the planar routing and pin redistribution. Since the emphasis is on completing vertical connections, the routing grid is divided into approximately ten vertical strips, and maze routing is performed within each strip on two layers to reduce memory requirements.

4. **Jog removal:** The planar routing procedure may generate a number of unnecessary bends or "jogs" in the wires. These lead to increased wire length and signal delays. The unnecessary jogs are removed during a clean-up phase, using a jog-removal algorithm based on plane sweeping [83]. Experimental results show that this step eliminates almost half the jogs, on the average.

5. **Routing area rotation:** As mentioned above, the division into columns biases the routing procedure in favor of horizontal nets. To reduce this bias, the routing grid is rotated by 90° after each layer, so that layers are alternately biased in the horizontal and vertical directions.

The planar routing step will be described here; details of the other steps can be found in [21]. The routing grid is divided by "columns," which are vertical grid lines containing at least one net terminal, into vertical channels. The planar

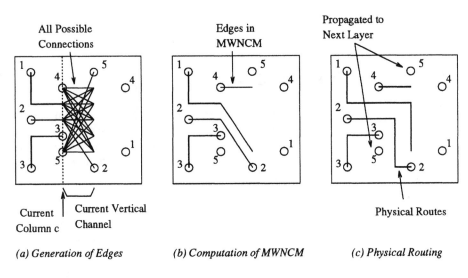

(a) Generation of Edges   *(b) Computation of MWNCM*   *(c) Physical Routing*

Figure 4.5   Planar routing steps in SLICE (after [21])

routing step consists of a series of channel routing steps performed in these vertical channels, starting from the leftmost channel. For a channel extending between the x-coordinates x_l and x_r (the coordinates are in grid units), the *channel capacity* C_{cap} is defined to be $x_r - x_l$. The capacity of a channel is the maximum number of vertical wires which can be accommodated in the channel. Note that due to the uneven distribution of net terminals on the top layer, the channels may differ in capacity. In each channel C_k, the left column contains a set of "start-points" which are either net terminals or endpoints of partial routes constructed in the channel C_{k-1} to the left of C_k. For each start point n_i, there is a corresponding *target* point *target*(n_i) to the right of n_i, which may or may not lie on the right column of the channel. The target point is unique, since all nets are assumed to be two-terminal nets (multiterminal nets are decomposed into two-terminal nets based on the minimum spanning tree).

For each channel, the planar routing procedure consists of three steps (Fig. 4.5):

1. *Generation of weighted edges*

2. *Computation of maximum noncrossing matching*

3. *Physical routing*

These steps are described below.

4.3.2   Generation of weighted edges

This step generates a set S of weighted edges connecting the set of start points P_l on the left column to grid points on the right column. These edges represent the different connections possible between the left and right columns, while the weight on a particular edge represents the *gain* of including the corresponding connection in the planar routing.

In general, each start point n_i, specified by its position (x_i, y_i) and the net net_i it belongs to, can be connected to any grid point on the right column which is free (not a terminal) or a terminal of net_i. However, this will result in the generation of too many edges. A simple *range-reduction* heuristic is used to reduce the number of edges: only edges with right end-point $y-$coordinates in the interval $[y_{i-C_{cap}}, y_{i+C_{cap}}]$ are generated, where y_{i-n}, y_{i+n} refer to the $y-$coordinates of the nth start-point above and below the start-point n_i. The reasoning behind this heuristic is as follows: assume that every start-point in the left column gets connected to some grid point in the right column. Then the end-point of the edge from n_i cannot lie outside $[y_{i-C_{cap}}, y_{i+C_{cap}}]$ as this will violate the channel capacity constraint (Fig. 4.6).

The determination of edge weights is accomplished as follows. If the target node n_j for start node n_i lies on the right column, the weight of the edge (n_i, n_j) is set to a high value $weight_{completed}$, since inclusion of this edge in the final routing will be very desirable. Edges with end-point $y-$coordinates in the *preferred region* $[y_i, y_j]$ are assigned a relatively high weight (even if n_j does not lie on the right column), since inclusion of one of these edges does not increase the wire length of net_i. Edges with end-points outside the preferred region have smaller weights which decrease linearly as the end-points move farther from the preferred region (Fig. 4.7).

4.3.3   Computation of maximum noncrossing matching

After the generation of the set S of possible connections, the next step is to select a subset of edges such that the total weight of the selected edges is maximized, and all the edges are *noncrossing*. This problem is referred to as

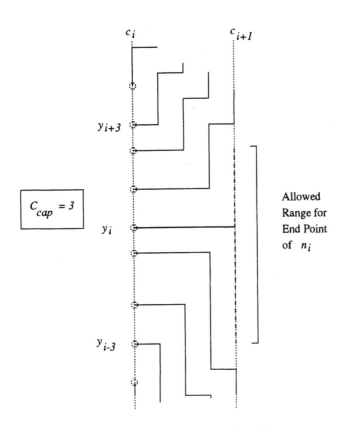

Figure 4.6   Range-reduction heuristic

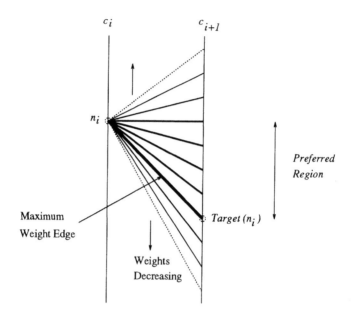

Figure 4.7   Edge weighting

the *maximum weighted noncrossing matching (MWNCM)* problem. This step is the core of the SLICE router. The effectiveness of SLICE relies on the fact that the MWNCM problem can be solved optimally in $O(n \log n)$ time, where n is the number of edges in the set S generated in the previous step.

The basic idea behind the MWNCM algorithm is the following: each edge in S is specified by a four-tuple (l, r, w, net), where l and r are the y-coordinates of the left and right ends of the edge, respectively, w is the weight of the edge and net is the net number. Each edge (l, r, w, net) is mapped to a unique point $(x, y) = (l, r)$ in the $x - y$ plane. The mapped points are assigned weights equal to the weights of the corresponding edges. Two types of *dominance* relationships are defined on the mapped points:

Strict dominance: A point $p_i = (x_i, y_i)$ *strictly dominates* another point $p_j = (x_j, y_j)$ if $x_i \geq x_j$ and $y_i > y_j$.

Lateral dominance: A point $p_i = (x_i, y_i)$ *laterally dominates* another point $p_j = (x_j, y_j)$ if $x_i \geq x_j$ and $y_i = y_j$ and $net_i = net_j$.

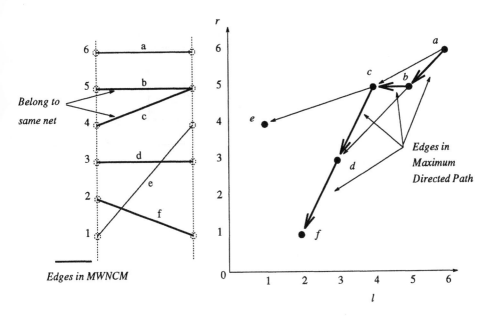

Figure 4.8 Mapping of edges to the $x - y$ plane

Thus, points related by a strict dominance relationship correspond to non-crossing edges, whereas points related by a lateral dominance relationship correspond to edges which belong to the same net and have the same end point. This mapping is illustrated in Fig. 4.8 (for clarity, some of the dominance relationships are not shown). The lateral dominance definition is introduced to allow the merging of edges belonging to subnets of the same multiterminal net into a Steiner tree.

It is intuitively clear that under this mapping, the MWNCM problem is equivalent to finding a subset of the mapped points, such that the total weight of points in the subset is maximized, and every point in the subset is related to every other point by a dominance relationship. This intuition is formalized in [21]. The new problem can be solved by constructing a *dominance graph*, where each node represents a mapped point, and a directed edge is introduced from p_i to p_j if p_i dominates p_j. The dominance graph can be shown to be a directed acyclic graph (DAG), and a maximum weighted path in the DAG corresponds to a solution to the MWNCM problem. This can be easily found in $O(n^2)$ time, where n is the number of vertices in the DAG (and hence the number of edges in S). However, the procedure can be made even more efficient using the

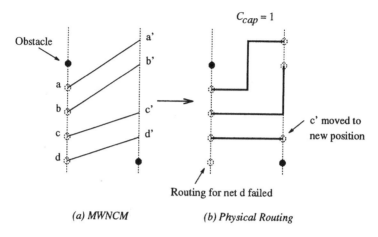

Figure 4.9   Physical routing from MWNCM

priority search tree data structure as described in [21], reducing the complexity to $O(n \log n)$.

4.3.4   Physical routing

The set of edges S_{MWNCM} gives a topological planar routing in the channel. This topological routing is converted to a physical (Manhattan) routing as follows: the edges in S_{MWNCM} are classified into *rising* edges and *falling* edges, depending on whether the end-point of the edge is above or below the start-point, respectively. The rising and falling edges do not interfere with each other (as they are noncrossing) and can thus be processed separately. The set S_{rise} of rising edges is sorted in decreasing order of y−coordinates, while the set S_{fall} of falling edges is sorted in increasing order. For edges in S_{rise}, physical routes are generated one at a time as follows: for each edge, the routing starts from the start point and advances vertically upward along the y−axis, until a blockage is reached, or the y−coordinate of the end-point is reached. Then, the routing advances one grid unit to the right (if possible) and repeats the process. The procedure ends when the route is either completed, or reaches the right column before reaching the end-point, or fails to reach the right column. In the first case, the completed route is recorded. In the second case, the partial route is recorded and the new end-point on the right column is propagated to the next layer. In the third case, the partial route is removed and the start-point and target point are propagated to the next layer (Fig. 4.9).

In experimental comparisons, the SLICE router was able to reduce the number of vias significantly in comparison to a 3-D maze router. The memory and CPU requirements were also much less for SLICE. One industrial example was too large to be handled by a 3-D maze router; however, SLICE was able to complete the routing using reasonable computing resources. However, the maze router was able to achieve smaller total wire length in fewer routing layers than SLICE. This is probably due to the greedy layer-by-layer approach taken by SLICE. A more global outlook may be able to improve the results significantly.

4.4 FOUR VIA ROUTING

In SLICE, due to the propagation of terminals from partial routes, a large number of vias may be generated. Although the overall via usage was shown experimentally to be significantly lower than that of a maze router, the number of vias generated in a particular net may be large. This can cause timing problems, since a net with many vias has greater capacitance and discontinuities in its transmission line impedance, which cannot be predicted before the routing. It is desirable to have a router which can *guarantee* that each net will have no more than a certain number of vias, so that a worst-case timing analysis can be performed before the routing, taking into account the effect of the vias.

In [81], a general-area multilayer MCM router "V4R" is described which uses no more than four vias for routing each two-terminal net, and no more than $4(k - 1)$ vias for each $k-$terminal net. The routing approach is similar to that of SLICE: division of the routing grid into slices, and proceeding layer by layer. However, instead of operating on one layer at a time, V4R operates on "x-y plane pairs," of which one layer is used for horizontal wires and the other layer for vertical wires. Under this routing model, which is sometimes called the *unidirectional layers* model, each bend in a net introduces a via between the layers of the plane pair. Such vias are called *secondary vias*, whereas vias connecting wire segments to net terminals are called *primary vias* [24]. The V4R router guarantees that every two-terminal connection will be routed using no more than four bends, or equivalently, using no more than five wire segments. Thus each connection will have at most four secondary vias.

The reason for allowing each net to have up to four vias is the following: if k vias (bends) are allowed in a net with linear dimension (length+width) n, the number of possible routes for the net is approximately $O(n^{k-1})$. Thus, if only one bend is allowed, we have only a constant number of choices (the two

L-shaped routes), whereas if we allow two bends, we have $O(n)$ choices (all possible Z-shaped routes), and so on. Increasing the number of possible route shapes gives greater routing flexibility and increases the chances of completing the routing. The minimum number of vias required to allow a route to be *non-monotone* (to cross a given horizontal or vertical line more than once) within the net bounding box is four. Non-monotone routes can increase completion rates at the expense of additional wire length, by their ability to go around obstacles. Increasing the number of vias beyond four yields marginal advantages, while reducing performance.

The V4R approach also eliminates the need for the restricted maze routing step required in SLICE. This reduces the memory requirement and increases the speed even further.

4.4.1   Overview of V4R

V4R routes the MCM substrate one $x-y$ plane pair at a time, starting from the top. As in SLICE, $k-$terminal nets are decomposed into $k-1$ two-terminal nets, using a minimum spanning tree algorithm. The two-layer 2-D routing grid is divided into *vertical channels* by vertical *columns*, where a column is a vertical grid line containing at least one net terminal. Similarly, horizontal *rows* divide the grid into *horizontal channels* (Fig. 4.10). On each plane pair, V4R processes columns one at a time, starting from the leftmost column. In each column, it generates a detailed physical routing directly, without the intermediate stage of global or topological routing.

When the rightmost column is reached, any incomplete routes are ripped up. Unrouted nets are then propagated to the next plane pair, and the process is repeated. The scan direction in the next plane pair is from right to left, to reduce the bias towards nets located on one side of the MCM. Unlike SLICE, there is no restricted maze routing step, and the grid is not rotated by 90° after each layer. Partial routes, which are allowed in SLICE, are not allowed here as they will cause an increase in the number of vias. Thus each net is forced to be routed completely in exactly one plane pair.

4.4.2   Planar routing in V4R

The major difference between SLICE and V4R is in the planar routing step. There is no topological routing step in V4R: the physical routing is generated

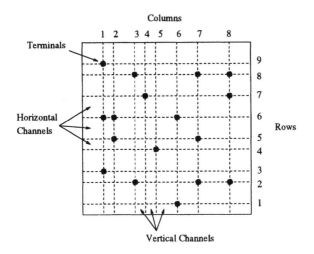

Figure 4.10 Division into rows and columns

directly. Each net i is assumed to be a two-terminal net, with the *left terminal* (terminal with the smaller x-coordinate) denoted by p_i and the right terminal denoted by q_i. For any terminal p, $x(p)$ and $y(p)$ denote the $x-$ and $y-$ grid coordinates of p, and $row(p)$ and $col(p)$ denote the row and column numbers of p. Every net is routed using one of two types of four-bend routing topologies: Type-1 topologies have three vertical segments (the left and right *v-stubs* and the *main v-segment*) and two horizontal segments (the left and right *h-segments*), while Type-2 topologies have three horizontal segments (the left and right *h-stubs* and the *main h-segment*) and two vertical segments (the left and right *v-segments*), as shown in Fig. 4.11.

In each column c, V4R executes the following steps:

1. **Horizontal track assignment of the right terminals:** In this step, for each right terminal q_i corresponding to a left terminal p_i in column c, V4R tries to find a horizontal track which is unoccupied between columns c and $col(q_i)$. For the q_i's which are successfully assigned to horizontal tracks, the nets will be routed using a Type-1 topology. These q_is are connected to their horizontal tracks by right v-stubs. The remaining nets will be routed using Type-2 topologies, with a right h-stub in $row(q_i)$.

2. **Horizontal track assignment of the left terminals:** In this step, for each net i which is to be connected with a Type-1 topology, V4R attempts

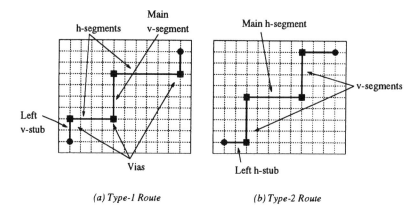

(a) Type-1 Route   *(b) Type-2 Route*

Figure 4.11   Four-via routing topologies

to assign the left terminal p_i to an unoccupied horizontal track for the left h-segment, which can be connected to p_i by a left v-stub. For nets with Type-2 topologies, it tries to assign a suitable horizontal track which will be used by the main h-segment.

For both types of nets, if the track assignment in step 2 fails, all wires for the failed net are ripped up and the net is added to the list of nets propagated to the next layer. If the assignment is successful, the net is called *active* if its routing has not been completed at the end of step 2. Active Type-1 nets need a main v-segment to complete the routing, while active Type-2 nets need the left and right v-segments. Main v-segments and left v-segments are called *pending* v-segments, while a right v-segment is called pending if the left v-segment of the net has been routed already.

3. **Routing in the vertical channel:** In this step, the router selects a set of pending v-segments and routes them in the channel between columns c and $c + 1$, subject to the channel capacity constraints.

4. **Extension to the next column:** The h-segments of remaining active nets are then extended to column $c + 1$. If the h-segment of net i cannot be extended due to obstacles, the remaining segments of i are ripped up and i is added to the list of nets for the next plane pair.

The first step is solved using an algorithm for computing a maximum weighted matching. A bipartite graph is constructed from the column instance as follows: each right terminal q_i corresponding to a p_i in column c is represented by a vertex on one side of the graph, and each horizontal track t_j which is free in

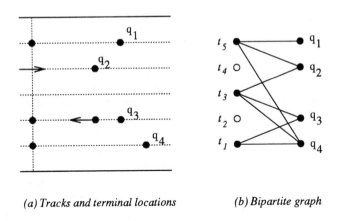

(a) Tracks and terminal locations   (b) Bipartite graph

Figure 4.12   Construction of bipartite graph

channel c is represented by a vertex on the other side of the graph (Fig. 4.12). Each vertex q_i is connected by an edge to every track t_j which is *feasible* for q_i. (A track is feasible for a terminal q_i if it is free between c and $col(q_i)$, or if it is occupied by a left or right terminal of net i.) The edges may have weights to reflect net priorities or other objectives. A *maximum weighted matching* in this graph corresponds to an optimal track assignment for the right terminals.

Note that the matching does not have to be a noncrossing matching, since the q_is may be in different columns. In case two terminals q_i and q_j are in the same column, then of all the horizontal tracks between q_i and q_j, the tracks closer to q_i are considered feasible for q_i, and the tracks closer to q_j are considered to be feasible for q_j. This avoids intersection of the right v-stubs of i and j, but may lead to nonoptimality of the solution.

The second step is solved in two phases. In the first phase, horizontal track assignment for the left terminals of Type-1 nets is done. A bipartite graph is constructed in a manner similar to the construction in the first step. In this case, however, all the p_is are in the same column c, so the matching corresponding to the track assignment must be noncrossing, to prevent intersection between the left v-stubs of the nets. This problem is solved using the MWNCM algorithm used in SLICE. In the second phase, feasible tracks for the main h-segments of Type-2 nets are assigned, again using a maximum weighted matching algorithm.

The next step is vertical channel routing. The objective in this step is to route as many pending v-segments in the current channel, subject to the channel capacity constraint. In general, the v-segments may be weighted, so the objective is to maximize the weight of the selected v-segments.

Each pending v-segment defines an *interval* $I = [a, b]$, where a and b are the row numbers of the end points of the segment and $b > a$. The pending v-segments are defined in such a way that no two of them have end points in the same horizontal track (this is referred to as a *lack of vertical constraints* in [81]). As a result of this definition, the only constraint on the routability of a set S of segments in a single vertical track is that the intervals I_1 and I_2 of any pair of segments $s_1, s_2 \in S$ should be non-overlapping if the segments belong to different nets. Thus, the channel routing problem is intuitively equivalent to finding k disjoint subsets of segments, where k is the channel capacity, such that the segments in each subset have non-overlapping intervals.

The problem is solved by defining a *partial ordering* relation on the set of segments, as follows: for two intervals $I_1 = [a_1, b_1]$ and $I_2 = [a_2, b_2]$, I_1 is defined to be *below* I_2, written as $I_1 \leftarrow I_2$, if (i) $b_1 < a_2$, or (ii) if $a_1 < a_2$ and $b_1 < b_2$ and both segments belong to the same net. This relation is shown graphically in Fig. 4.13: each interval is represented by a vertex, and a directed edge joins I_i to I_j if $I_j \leftarrow I_i$. Intuitively, the problem now is to find a set of k disjoint directed paths in the graph, such that the total weight of the vertices in the paths is maximized. An algorithm for finding a *maximum weighted k-cofamily* in a partially- ordered set (*poset*) [84] is used to solve the problem optimally in $O(km^2)$ time, where k is the channel capacity and m is the number of active nets crossing the channel.

The V4R router is able to achieve significant improvements over a 3-D maze router. In the experimental results presented in [81], V4R used 44% fewer vias and 2% less total wire length than the maze router, while running 26 times faster and using equal or fewer routing layers. These results are also much better than those of the SLICE router. Furthermore, the memory usage in V4R is minimal, since even the restricted maze routing step of SLICE is eliminated. The memory requirement scales linearly with decreasing line pitch, whereas the requirements of the maze router and SLICE scale quadratically. Thus, V4R is better suited for increasingly dense designs.

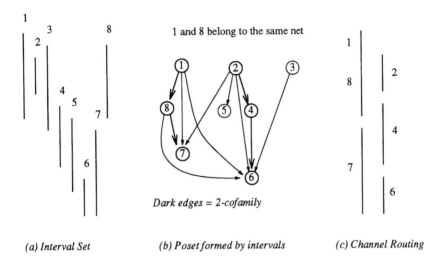

(a) *Interval Set*   (b) *Poset formed by intervals*   (c) *Channel Routing*

Figure 4.13   Vertical channel routing

4.5 RUBBER-BAND ROUTING

Grid-based routing schemes are inefficient for large, complex MCM designs. The SURF routing system [85] employs the concept of *rubber-band sketches* to represent MCM routing topologies efficiently without an underlying grid.

A *sketch* is a planar topological routing, consisting of a set F of *features*, connected by a set W of *wires*. Each feature is usually represented by a point. Each wire connects a pair of features, and no wire intersects itself or any other wire (Fig. 4.14(a)). There are many ways in which a topological routing can be represented by a sketch. One canonical form of sketch is the *rubber-band equivalent* of a sketch, introduced in [20], referred to as a rubber-band sketch in [86]. In a rubber-band sketch, each wire is represented by a rubber-band, which is a set of straight line segments, whose endpoints are in F (Fig. 4.14(b)). The rubber-band for a wire gives a minimum-length routing corresponding to the wire's topology.

The rubber-band routing procedure for an MCM consists of two stages: the creation of a multilayer rubber-band sketch and conversion of the sketch to a physical routing. The first step is accomplished by a topological router described in [87]. The input to the router consists of a set of terminals, a set of nets, a set of obstacles and a set of wiring rules. The wiring rules may include geometrical design rules and topological constraints on the nets, such as spe-

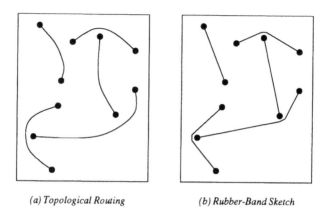

(a) Topological Routing   (b) Rubber-Band Sketch

Figure 4.14   A rubber-band sketch (after [86])

cific net topologies and bounds on wire lengths. The output of the router is a multilayer rubber-band sketch which is very likely to be routable on every layer, although the routability is not guaranteed.

The topological routing proceeds in two steps: global routing followed by local routing. In the global routing stage, the routing area is recursively partitioned into bins, until the sizes of the bins are small enough to be handled by the local router. During partitioning, the crossing points are specified on the bin interfaces for nets which are cut by the interface. The *least commitment principle* is used to avoid suboptimal solutions, by allowing the crossing points to move around on the interface during the partitioning procedure. Also, if a bin becomes too dense at some level of the partitioning, the problem is corrected by allowing the procedure to backtrack.

After partitioning is complete, the local router can be applied to route the nets sequentially in each bin. The local router finds topological paths connecting each pair of points in a *region graph*. The region graph represents the regions in the neighborhoods of the features, and their adjacencies. For multilayer routing, the region graph represents regions on all layers, with regions on different layers being defined as adjacent if they can be connected by a via. The computation of the shortest path in the region graph is accomplished efficiently based on an underlying data representation known as *constrained Delauney triangulation* [88]. The topological routes generated during this stage are guaranteed to be planar; however, they may not be valid rubber-bands as they may not be minimum-length. The routes are converted to valid rubber-bands during a validation step. Other heuristics for avoiding local congestion are presented

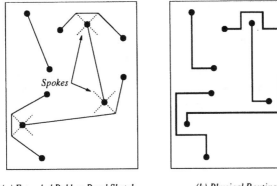

(a) Extended Rubber-Band Sketch   (b) Physical Routing

Figure 4.15   Extended rubber-band sketch and physical routing(after [86])

in [87]. These maximize the likelihood that the rubber-band sketch can be successfully transformed into a physical routing.

To convert a rubber-band sketch into a physical routing, the rubber-bands have to be "pulled away" from the features and from each other, so as to satisfy the minimum spacing design rules. This can be done by the addition of "spokes" at each feature, as described in [20]. For Manhattan wiring, each feature may have up to four spokes, at slopes of 45° and 135°. An *extended rubber-band sketch* is created when the spokes are added to the features, as shown in Fig. 4.15(a). A rubber-band sketch is *routable*, i.e., convertible to a physical routing, if and only if it can be transformed into an extended rubber-band sketch [86]. Figure 4.15(b) shows the physical routing derived from the extended rubber-band sketch.

The rubber-band representation has a number of advantages. It can be updated incrementally when a small change is made in the design. The representation provides reasonably accurate estimates of wire lengths, which can be used for timing verification before the final routing is completed, thus reducing design cycle time. The routability of a rubber-band sketch can be tested efficiently without performing an actual physical routing [86]. Thus, the rubber-band representation can be used until the design is finalized, after which it can be converted to a valid physical routing.

4.6 SUMMARY

This chapter gave an overview of the various constraints and objectives associated with the multilayer MCM routing problem. The problem is significantly different from PCB and VLSI routing problems, so completely new routing approaches have to be developed to solve it. Some of the popular recent algorithms developed specifically for high-performance multilayer MCM routing were surveyed.

5

PERFORMANCE-ORIENTED TREE CONSTRUCTION

As described in the previous chapter, multilayer MCM routing is a complex three-dimensional problem. A common approach to many difficult problems is the "divide-and-conquer" strategy, and this can be effectively applied to the multilayer routing problem also. The three-dimensional routing problem is decomposed into two subproblems: one solves the $x-$ and $y-$dimensions, and the other solves the $z-$ dimension. The first subproblem is usually called *planar* or *single-layer* routing, and the second one is called *layer assignment*. There are at least two different routing approaches based on this partitioning of the problem, depending on the order in which the subproblems are solved:

1. *Single-layer routing is performed first, followed by layer assignment.* Initially, two-dimensional route shapes are determined for all the nets. These route shapes may intersect or overlap with each other, causing a nonplanar situation. Later, these conflicts are resolved during the layer assignment stage, by assigning conflicting routes (or parts of routes) to different layers, so as to obtain a planar set of routes on every layer. Following the terminology of VLSI routing, this approach may be called *constrained* multilayer routing [89], since the route shapes are fixed during the layer assignment phase.

2. *Layer assignment is performed first, followed by planar routing.* Initially, nets are assigned to a specified number of layers, with the objective of maximizing the routability of nets on each layer. After layer assignment, single-layer routing is performed on each layer. This approach is called *unconstrained* or *topological* multilayer routing [90, 91].

For high-performance designs, the first approach presents some significant advantages. As was shown in Chapter 2, the delay in an interchip net is a strong function of the net topology. The constrained multilayer routing approach gives us the maximum control over net topologies, since they are fixed in the first step. In the other approaches, the constraint of planarity may lead to a suboptimal net topology.

Second, since the net route shapes are known before layer assignment, it is easier to estimate effects such as crosstalk between pairs of nets. Strongly interfering nets can then be placed on different layers to minimize crosstalk. This issue is discussed in greater detail in Chapter 6.

The knowledge of the route shape also allows us to estimate the *timing criticality* of a net before layer assignment. If a net is highly critical, the layer assignment process can be instructed to assign the net to a layer as close to the chip layer as possible, and to avoid splitting the net across different layers. Minimizing the number of vias in a net leads to better performance, since the vias introduce large parasitic capacitances and discontinuities in the transmission lines.

This chapter discusses the single-layer routing problem, which forms the first step of the constrained multilayer routing process. Constrained and unconstrained layer assignment are discussed in the next chapter.

5.1 PERFORMANCE-DRIVEN TREE GENERATION

The goal of the first stage in constrained multilayer routing is to generate, for each net, a tree connecting all the terminals of the net. The terminals are assumed to lie on a regular rectangular grid. At this stage, the routes need not be planar, i.e., trees of different nets are allowed to intersect and overlap.

The problem of generating trees for multiterminal nets has been studied extensively. The most common objective for tree generation has been wire length minimization. There are two reasons for this: first, in VLSI design, reducing wire length reduces the overall chip area, thus increasing the yield of the chip. Second, under the lumped capacitor delay model, net delay is directly proportional to wire length, hence reducing wire length leads to faster chips. A large portion of the work in VLSI global routing has been focused on constructing minimum-length Steiner trees. Only recently, timing-driven Steiner tree generation algorithms have begun to emerge [92, 93, 94].

In MCMs, the resistance and inductance of the interconnect cannot be ignored, due to the smaller driver impedance values and greater lengths of the wires, as compared to on-chip interconnections. Depending on the speed of operation and the relative values of the electrical parameters R, L and C of the interconnect and driver, more sophisticated delay models such as those described in Chapter 2 may be necessary. Under these delay models, minimum wire length does not necessarily correspond to minimum delay.

Sections 5.2 and 5.3 describe previous approaches to tree generation, based on the first-order Elmore and dominant time constant delay models, respectively. Section 5.4 presents a new tree construction approach based on the second-order delay model of Section 2.9.2 and a consideration of reflection effects.

In addition to interchip signal delays, *clock skew* is another important factor which limits system clock speeds. Typically, a clock signal distribution tree has a large number of fanouts or sinks. The clock skew is defined as the maximum difference in arrival times of the clock pulse at any of the sink pins. To maximize system speed and guarantee proper operation of a synchronous system, it is desirable to minimize clock skew. Section 5.6 describes approaches to construct clock trees with minimal clock skews.

5.2 TREE CONSTRUCTION BASED ON ELMORE DELAY

The Elmore delay model has been shown to accurately model physical delay in certain situations: optimizing Elmore delays in 4- and 5-pin signal nets is shown in [95] to produce near-optimal routing trees, as verified by Spice simulations. One of the first global routing algorithms which considered interconnect resistance in the delay expression was presented in [96]. The global routing algorithm first finds trees for all the nets using the A^*-search algorithm [97], such that the maximum delay to any sink node in the tree is minimized. In the second stage, the algorithm iteratively rips up and reroutes less critical nets, so as to reduce congestion.

The A^*-search algorithm is a graph-searching heuristic, which can be used to find a minimum cost path in a graph between two vertices. In [96], it is used to

```
ERT(N = {n_0, n_1, ..., n_k}) {
    T = (V, E) = ({n_0}, φ);
    while |V| < k do {
        find best u ∈ V and v ∉ V;
        V ← V ∪ {v};
        E ← E ∪ {(u, v)};
    }
    output spanning tree T(V, E);
}
```

Figure 5.1   Elmore routing tree algorithm

find a path in a routing graph, from a partially constructed tree T to a target vertex z^*. The algorithm assigns a value f to each vertex being considered as part of the path. The value f is a measure of the benefit, or estimated benefit, of a path passing through the vertex. The benefit is computed based on the estimated Elmore delay from the source to all sink pins, and the total net length. The algorithm terminates when it reaches the target vertex, and backtracks along the path which maximizes the f value.

A routing tree is constructed by initializing T to the source vertex n_0 of the tree, and then finding optimal paths from T to each sink node $n_i, i = 1, ..., k$ sequentially, adding the paths to T at each step.

The average complexity of the A^*-search algorithm is $O(v)$, where v is the number of vertices in the routing graph. Thus the overall complexity of this procedure is $O(mnk^2)$ where $m \times n$ is the size of the routing grid, and k is the number of sinks in the net.

Another class of greedy *Elmore routing tree* (ERT) algorithms is presented in [98]. These algorithms also attempt to optimize Elmore delays directly during the construction of the routing tree. The first algorithm in this class is called the ERT algorithm. It constructs a *spanning tree* which minimizes the maximum Elmore delay at the sink nodes. The outline of the algorithm is shown in Fig. 5.1.

As in [96], the ERT algorithm begins with the trivial tree T consisting of the source node n_0. In each step, the algorithm finds two sink nodes $u \in T$ and $v \notin T$, such that adding an edge (u, v) to T will minimize the maximum Elmore delay to any sink node in $T \cup v$. Unlike [96], where the order in which the sink nodes are added is fixed, the ERT algorithm finds the best new vertex to add at each step. It is possible to construct problem instances in which the delay of the tree constructed using the algorithm of [96] is at least twice as large as the Elmore delay of the tree constructed using the ERT algorithm.

The ERT algorithm generates only spanning trees since only edges of the form n_i, n_j are allowed in the tree. It can be generalized to a *Steiner Elmore routing tree* (SERT) algorithm, in which the new sink node is allowed to be connected to an *edge* of the partial tree T. The edge of T is then split into two edges, meeting at a Steiner point w which is closest to the new sink V, and a new edge (v, w) is added to T.

While the A^* and ERT algorithms attempt to minimize Elmore delay directly, a number of other algorithms have been proposed based on intuitions derived from the delay model. Examination of the Elmore delay model reveals that the total delay is related linearly to the total wire length as well as quadratically to the source-sink path lengths. Thus, intuitively, a good tree should have small overall wire length as well as a small "radius," where the radius of a tree is defined to be the maximum path length from the source to any sink node. A *bounded radius* Steiner tree algorithm is presented in [99], which constructs trees such that the total wire length is at most $2(1 + 2/\epsilon)$ of the optimal Steiner tree wire length, and the longest path is at most a factor of $(1 + \epsilon)$ times the maximum source-sink Manhattan distance. A *performance-oriented minimum rectilinear Steiner tree* (POMRST) is presented in [100] which minimizes total wire length subject to constraints on the maximum lengths of all source-sink paths. An expression for an upper bound on the Elmore delay is used to guide two performance-driven Steiner tree algorithms in [101].

5.3   GLOBAL ROUTING USING A-TREES

An elegant solution to the tree-construction problem based on the dominant time constant delay model is described in [74]. The starting point for their algorithm is a decomposition of the expression for delay into four components, as described in Section 3.3.1. The expressions are repeated here for convenience:

$$t_d(T) = \sum_{k \in T} C_k R_{p(d,k)} \qquad (5.1)$$

This expression can be split into four terms and rewritten as

$$t_d(T) = t_1(T) + t_2(T) + t_3(T) + t_4(T) \qquad (5.2)$$

where

$$
\begin{aligned}
t_1(T) &= R_d \sum_{k \in T} C_0 \\
t_2(T) &= R_0 \sum_{k \in sink(T)} l_{p(d,k)} C_k \\
t_3(T) &= R_0 C_0 \sum_{k \in T} l_{p(d,k)} \\
t_4(T) &= R_d \sum_{k \in sink(T)} C_k = constant
\end{aligned}
$$

where R_0 and C_0 are the resistance and capacitance per unit length of the wire, R_d is the driver resistance, C_k is the gate capacitance at sink node k, and $l_{p(d,k)}$ is the length of the path from the driver to node k.

The first term, $t_1(T)$, is proportional to the total wire length of the tree. Thus, when R_d and C_0 dominate the other parameters, a minimum wire length Steiner tree is desirable. The second term is proportional to the sum of driver-to-sink path distances, and becomes significant when the wire resistance and sink capacitances become large. The third term is related quadratically to the path lengths in the tree, and this also becomes significant when wire resistance increases. Since the fourth term is a constant, it can be ignored for optimization purposes.

Thus, there are four *time constants*, $R_d C_0$, $R_0 C_k$, $R_0 C_0$ and $R_d C_k$, the fourth of which may be ignored. For a tree to have minimum delay in the general case, when no single time constant dominates, the sum of path lengths from driver to sinks must be kept as short as possible, while simultaneously minimizing the total wire length and the maximum path length. A new type of tree, called the *A-tree*, was shown in [74] to have some unique properties with respect to these cost functions.

By definition, an A-tree is a tree in which every node is connected to the driver by a shortest length path. Thus, $t_2(T)$ is always minimum. Further, it is shown in [74] that when $t_1(T)$ is minimized in an A-tree, $t_3(T)$ is simultaneously minimized. Thus the problem of simultaneously minimizing t_1, t_2 and t_3 reduces to finding a minimum-length A-tree.

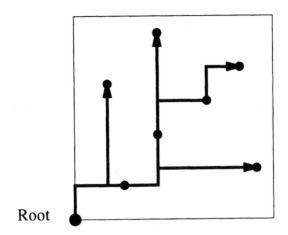

Figure 5.2   Rectilinear Steiner arborescences

A-trees are a generalization of *rectilinear Steiner arborescences (RSAs)* [102].
An RSA is a directed version of a Steiner tree, in which all vertices lie in the
first quadrant, with the root at the origin. All horizontal edges are directed
from left to right, and all vertical edges from bottom to top (Fig. 5.2). The
A-tree is an extension of the RSA to four quadrants: the root remains at the
origin, but the other vertices may be located anywhere. All edges must be
directed away from the root (Fig. 5.3). Thus, an A-tree consists of four RSAs,
one in each quadrant, with a common root.

It is not known yet whether the problem of constructing a minimum-length RSA
(*RMSA*) is NP-complete. An approximation algorithm for A-tree construction
was presented in [102], which guarantees an RSA no more than twice as long
as the RMSA. Another algorithm is presented in [74], which performs very well
in practice. The algorithm constructs a tree using a set of "safe" moves as far
as possible, and uses a heuristic move whenever a safe move is not possible. It
is shown that a tree constructed using only safe moves is optimal. In practice,
96% of the moves used by the algorithm are safe moves, and the constructed
trees are nearly optimal in terms of total length.

The A-trees were compared to trees generated by a Steiner tree algorithm.
The source-to-sink delays were significantly lower for the A-trees (as much
as 45% lower for 16-terminal nets). This shows that A-trees are much better
than Steiner trees for performance-driven routing when wire resistance becomes
significant.

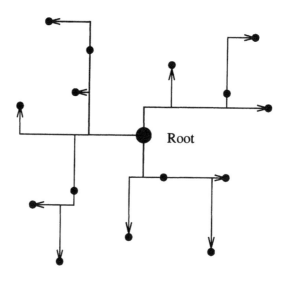

Figure 5.3   Example of an A-tree

5.4   TREE GENERATION USING THE RLC DELAY MODEL

As discussed in Chapter 2, RC delay models are inadequate when transmission line effects become significant, as in MCM interconnects. In such situations, a second-order delay model such as the one presented in Section 2.9 is more appropriate for modeling the effects of mismatch of the driver and transmission line impedance.

Unfortunately, the second-order delay model is not as amenable to analytical optimization as the first-order models, due to the complexity of the expressions for the second-order parameters b_1 and b_2 and the lack of an analytical delay expression. The second-order delay expression of Eq. (2.89) is repeated below:

$$t_d = \begin{cases} \dfrac{1}{1 - 0.93745e^{-2.62321K^{-0.68}}}b_1 & K \leq 4\pi^2 + 1 \\[2ex] \dfrac{K}{2}b_1 & K > 4\pi^2 + 1 \end{cases} \qquad (5.3)$$

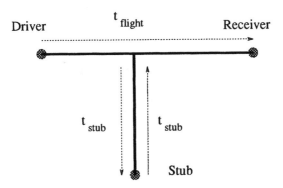

Figure 5.4   Effect of a stub on signal delay

This expression indicates that for delay minimization under the second-order delay model also, it is important to minimize b_1. Thus, A-trees can be expected to perform well even under this delay model. The following sections describe an A-tree construction algorithm which is guided by the second-order delay model and attempts to minimize the impact of reflections from unterminated stubs.

5.4.1   Stub minimization

A *stub* or branch in a tree introduces extra delay and/or ringing in the received signal waveform. When the wavefront from the driver reaches a branch point, it splits into two waves, each of smaller amplitude than the original. One wave continues down the main path, while the other one travels down the stub, gets reflected from the end of the stub, comes back to the branch point, and eventually reaches the receiver node R (Fig. 5.4). The waveform at R does not reach its full value until the reflection from the stub arrives at R. Thus the stub adds an extra delay equal to twice the time of flight of the signal across the stub.

Ideally, eliminating stubs entirely from the tree would result in the "cleanest" waveform at all the receiver nodes. This is called "daisy-chaining" (Fig. 5.5). In general, however, daisy-chaining carries an unacceptably high wire length penalty, resulting in higher power dissipation with no appreciable gain in performance.

In our tree-generation algorithm (Fig. 5.6), the following approach is taken: in each quadrant, an *x-y monotone* path containing the source node and the

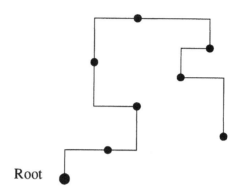

Root

Figure 5.5   Daisy-chaining of sink nodes

$T := \phi$;
for each quadrant $(Q_i, i = 1, \ldots, 4)$ {
    $T := T \cup$ maximal monotone path (Q_i);
    for each sink $(s \in Q_i)$ {
        $T := T \cup$ optimal path (s, T);
    }
}

Figure 5.6   Tree-generation algorithm

maximum number of sink nodes is constructed initially, in order to minimize the number of stubs. (An $x - y$ monotone path from (x_1, y_1) to (x_2, y_2) is a path along which the $x-$ and $y-$ coordinates increase or decrease monotonically, from x_1 to x_2 and from y_1 to y_2. Equivalently, it is a *minimum Manhattan distance path* between the two points.) This can be done optimally, as described below. The sink nodes which do not lie on this path are then attached to the partial tree to minimize the second-order delay as described in the next section. To simplify the discussion, we restrict our attention here to the first quadrant.

A point (x_1, y_1) is said to *dominate* another point (x_2, y_2) if $x_1 \geq x_2$ *and* $y_1 \geq y_2$. To find the x-y monotone path containing the maximum number of sink nodes, a directed acyclic graph (DAG) is constructed with the source and sink nodes as the vertex set as follows: a directed edge is introduced from vertex u to v if and only if v dominates u. The set of vertices dominated by v is called the *ancestor set of* v, denoted by A_v. Each vertex v is associated with an index d_v, which is initialized to zero. The vertex set is then scanned

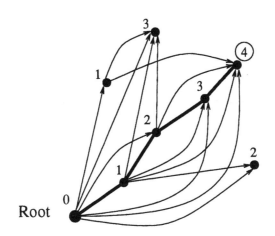

Figure 5.7   Computation of maximal monotone path

repeatedly. In each scan, the index of a vertex v is updated to 1 greater than the maximum index of any vertex $u \in A_v$. The scanning continues until the maximum index value does not change further. Figure 5.7 shows an example with the final index values for each vertex.

At this point, the x-y monotone path passing through the maximum number of sink nodes can be found by backtracking from the vertex v having the maximum d_v, choosing the ancestor with the largest index at each step. The bold line in Fig. 5.7 shows the maximal path for this example.

For each edge in the path, there may be several rectilinear realizations. It is usually desirable to minimize the number of bends in the tree, since each bend introduces a via if the substrate layers are restricted to unidirectional wiring. Bends also introduce discontinuities in the transmission lines, increasing delay. Each edge in the path can be realized using at most two different single-bend connections. A Type-1 connection consists of a vertical wire followed by a horizontal wire; the order is reversed in a Type-2 connection. The choice between a Type-1 and a Type-2 connection is made as follows: consider an edge joining two nodes $a(x_1, y_1)$ and $b(x_2, y_2)$, b dominates a. Let the set of sink nodes in the region $x_1 \leq x \leq x_2, y \geq y_2$ be V, and the set of sink nodes in the region $x \geq x_2, y_1 \leq y \leq y_2$ be H. If H is empty, a Type-1 connection is chosen; otherwise, if V is empty, a Type-2 connection is chosen. If neither is empty, we define $h = \min_{i \in H}(x_i - x_1)$ and $v = \min_{i \in V}(y_i - y_1)$. If $v > h$, a Type-1 connection is chosen; otherwise, a Type-2 connection is chosen. These

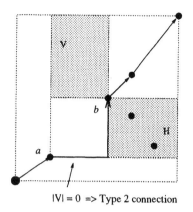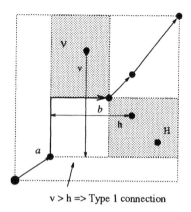

|V| = 0 => Type 2 connection   v > h => Type 1 connection

Figure 5.8   Two choices for a L-shaped connection

situations are illustrated in Fig. 5.8. These rules tend to minimize the length of stubs attached to the main path.

5.4.2   Attaching remaining nodes

After the maximal path (or "main path") is constructed, the admittance values at each point in the path are computed as described in Chapter 2. The main path forms the initial partial routing tree for the net. The sink nodes that remain unconnected after the main path is constructed are attached to the partial tree one at a time, in ascending order of their distance from the root. For each unconnected node u, an optimal point k is found in the partial tree, such that when u is attached to k, the maximum delay to u and all connected sinks is minimized.

The optimal point k is found as follows: first, the set R_u of points in the partial tree which are dominated by u, and are *reachable* from u, is identified. A point v is reachable from u if there exists an x-y monotone path from v to u which does not pass through any other vertex in the partial tree. Figure 5.9 shows an example of a partial tree and an unconnected vertex u. The shaded rectangle is the set of points dominated by u; the set $R(u)$ is shown with bold lines.

After R_u is constructed, the following steps are repeated for each point v in R_u:

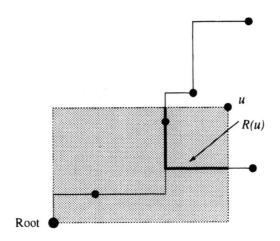

Figure 5.9 A partial tree and an unconnected node u

1. A "trial path" is introduced from v to u.

2. The admittance values for the nodes in the partial tree are updated.

3. The delays to all sink nodes in the partial tree are recomputed, and the maximum delay is stored.

4. The trial path is destroyed and the admittances are restored to their original values.

The optimal point k is the one for which the maximum sink delay is minimized. The node u is attached to the partial tree at node k, and the admittances in the partial tree are updated.

Updating the admittance values when a trial path is added (or removed) at a point v can be accomplished very efficiently. Note that only the admittance values of nodes "upstream" of v will be affected. The second-order admittance of a straight path (without branches) of length l can be computed analytically using the expressions derived in Section 2.7. Thus the extra admittance ΔY at node v can be computed very easily. The new admittances of nodes upstream of v can be computed incrementally as described in Section 2.5.4. When the admittance values have been updated, the new coefficient values to any connected sink node s can be found by backtracking from s to the driver node.

To remove the trial path, the same procedure is repeated for updating the coefficients, except that the sign of ΔY is reversed.

Table 5.1   Comparison of three tree construction algorithms

k	Max. Sink Delay (ns)			Total Wire Length		
	RLC	Elmore	Steiner [103]	RLC	Elmore	Steiner [103]
6	1.37	1.37 (+0%)	2.19 (+60%)	93	93 (+0%)	83 (-11%)
8	2.38	2.42 (+2%)	3.46 (+45%)	155	148 (-5%)	121 (-22%)
8	2.66	2.72 (+2%)	3.66 (+38%)	195	153 (-22%)	125 (-36%)
8	1.59	1.59 (+0%)	1.80 (+13%)	108	108 (+0%)	103 (-5%)
9	1.78	1.78 (+0%)	2.58 (+45%)	123	121 (-2%)	101 (-18%)
12	2.00	2.28 (+14%)	3.19 (+60%)	202	185 (-8%)	135 (-33%)
12	1.88	2.03 (+8%)	2.88 (+53%)	191	185 (-3%)	124 (-35%)
13	1.92	1.80 (-6%)	4.27 (+122%)	177	162 (-8%)	131 (-26%)
15	2.28	2.89 (+27%)	4.09 (+79%)	222	197 (-11%)	154 (-31%)
16	2.77	3.18 (+15%)	5.26 (+90%)	254	227 (-11%)	189 (-26%)
16	2.17	2.26 (+4%)	5.28 (+143%)	183	180 (-2%)	161 (-12%)
21	2.54	2.73 (+7%)	3.89 (+53%)	243	268 (+10%)	203 (-16%)
23	3.16	3.06 (-3%)	4.78 (+51%)	308	287 (-7%)	214 (-31%)
25	3.59	3.78 (+5%)	5.61 (+56%)	337	319 (-5%)	213 (-37%)

5.4.3   Experimental Results

The performance of the new second-order delay-based tree generation algorithm
was measured on a number of randomly generated multiterminal nets. The
results were compared with a high-quality minimum-wirelength Steiner tree
generation algorithm [103]. To measure the effectiveness of the second-order
delay model, the algorithm was also run using only the Elmore delay, i.e.,
ignoring the inductance and second-order effects.

A comparison of typical results using the three approaches is shown in Table 5.1.
The first column shows the number of terminals in the net. The second column
shows the delay of the tree constructed using the second-order delay model, in
nanoseconds. The third column shows the delay when the new algorithm was
guided by the Elmore delay model, and the last column shows the delay of the
tree constructed by the Steiner algorithm. In all cases, the delay was computed
using the second-order delay model. The delay of a tree is defined to be the
maximum source-sink delay among all sink nodes in the tree. The last three
columns show the wire length of the trees constructed using the three different
algorithms.

As the table shows, the tree construction algorithm using the second-order delay model is able to construct trees with significantly smaller delays than the Steiner tree algorithm. On the average, the delays in the Steiner trees were 65% greater, and in one case the delay was nearly two and a half times the delay in the tree constructed using the new algorithm. Using the second-order model almost always resulted in a smaller delay, as compared to using the Elmore model, and in one example, the delay was reduced by 27%.

Due to the reduction in the number of stubs, the waveforms at the sink nodes were also improved, as shown in Fig. 5.10. The figure shows waveforms at the same sink node of a net, when it was routed using a Steiner tree, and using the new algorithm.

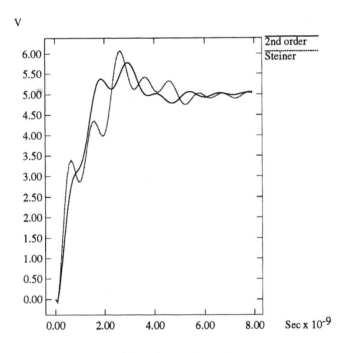

Figure 5.10   Sink node waveforms for two trees

The price to be paid for reduced delay is the longer total wire length. The Steiner trees were often significantly shorter than the A-trees. On the average, the Steiner trees were about 26% shorter than the A-trees constructed using the second-order model, and the A-trees using the Elmore model were about 5.6% shorter on the average. Longer wire length implies that more wiring resources (such as routing layers) may be required to complete the routing. In the case of thin-film MCMs, where the amount of wiring resources available, even on two layers, is often much more than required, this may not be a problem, but on laminated or ceramic MCMs, each additional layer increases the cost of the package. Also, longer wires may result in greater power dissipation, which again pushes up the cost of the package due to additional cooling requirements. Thus, the new tree construction algorithm is suitable for situations where

1. performance is very critical,

2. low power consumption is not essential, and

3. cost is not a primary criterion.

In situations requiring a compromise solution, an iterative modification algorithm can be applied, which begins with a fast but long tree, and shortens it iteratively while satisfying the given timing constraint.

The routing algorithm based on the second-order delay model is able to take into account delay increases caused by driver and line impedance mismatches. However, it is unable to account for oscillations caused by unterminated stubs in the tree. These oscillations cannot be predicted without using a higher-order model for the interconnect, which is computationally expensive during routing. An interesting extension to this work would be to derive approximate models for the effect of stubs, in terms of their termination capacitance and length, and use these models during routing to minimize delays and improve signal integrity by modifying the tree topology.

5.5 MINIMUM CONGESTION ROUTING

The tree construction algorithms discussed above construct the net trees sequentially. During the construction of a tree, the remaining nets in the netlist are ignored, since the overall global routing is not expected to be planar. Non-planar situations such as wire crossings and track sharing are resolved during the layer assignment stage. The advantage of this approach is that it frees the tree construction stage from physical constraints, so that the individual trees can be constructed to maximize performance. However, in the final multilayer routing, this approach may lead to poor performance. For example, ignoring the presence of other nets may lead to excessive crosstalk between nets, thus increasing noise levels and delays. A 2-D routing with local regions of high density may require more layers for successful completion of the routing, since a lower bound on the number of layers depends on the region with maximum density in the 2-D routing [24].

One way to avoid creating regions with high density is to consider *congestion* in the routing grid. The congestion of an edge in the routing grid is an estimate of the number of nets which will pass through that edge in the final 2-D routing. As nets gets routed one by one, the congestion information becomes more accurate, as the estimates are replaced by actual values. The tree construction algorithms have to be modified to minimize the total cost of all the edges in the tree, where the cost of each edge is related to the congestion of the edge. Thus, the algorithms will tend to avoid congested regions whenever possible.

There are many ways of estimating the congestion. The simplest approach is to record the number of nets passing through each edge, as the nets are routed. However, this ignores nets which have not yet been routed, making the effectiveness of the approach dependent on the order in which the nets are routed. To reduce the order dependence, it is necessary to estimate the congestion of unrouted nets. One way of doing this is described in [104]. It is assumed that each net will be routed using a single-bend (L-shape) route. The congestion in the grid is estimated by superimposing all possible L-shaped routes for all the nets on the grid, and then counting the number of routes passing through each grid edge.

The effectiveness of the congestion estimation approach depends on the routing algorithm used. For example, if the algorithm uses only single-bend routes, the approach of [104] is appropriate. If any $x-y$ monotone route is equally likely to be generated, or if the characteristics of the router are not known, the following estimation approach can be used. Consider a two-point connection, as shown in Fig. 5.11. The width and height of the enclosing box are n and m, respectively.

Assuming that all $x-y$ monotone paths are equally likely, the probability that a path will pass through a particular edge can be computed recursively. Assume that all vertical grid edges are directed from top to bottom, and all horizontal edges from left to right. The probability that the path will pass through a vertex $v_{i,j}$ is equal to the sum of the probabilities of the edges $(v_{i-1,j}, v_{i,j})$ and $(v_{i,j-1}, v_{ij})$. Let this probability be P_{ij}. It is easy to show that the ratio of the number of paths from v_{ij} to the terminal $v_{1,1}$ which pass through the vertical edge $(v_{i,j}, v_{i+1,j})$ and the horizontal edge $(v_{i,j}, v_{i,j+1})$ is given by $i :: j$. Thus, the probability of the path passing through the edge $(v_{i,j}, v_{i+1,j})$ is given by

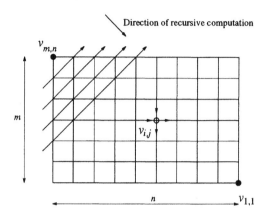

Figure 5.11   Congestion estimation for an unrouted net

$\frac{i}{i+j} P_{ij}$, and the probability of the edge $(v_{i,j}, v_{i,j+1})$ is given by $\frac{j}{i+j} P_{ij}$. The starting point for the recursion is $P_{mn} = 1$, since the path must pass through the terminal v_{mn}.

5.6   CLOCK TREE ROUTING

In a synchronous digital VLSI system, the operation of the functional blocks is synchronized by a clock signal, which is usually fed from a single clock buffer to all the clocked elements or latches by a large *clock tree*. The delay from the output of the buffer to each sink node of the clock tree may not be a constant, so that the clock signal reaches different sink nodes at different instants of time. The difference between the earliest and latest clock arrival times is termed the *clock skew*. To ensure correct operation of the system, the minimum clock period T_{clk} should satisfy the following relation:

$$T_{clk} \geq d_{max} + T_o + skew$$

where d_{max} is the maximum signal delay of any latch-to-latch path and T_o is a constant delay factor which takes into account internal delays and set-up times of latches and safety margins. In high-performance systems, d_{max} and T_o are relatively small, and the clock skew can have a large impact on the maximum achievable clock frequency.

Several clock tree routing algorithms have been proposed for minimizing or eliminating clock skew. A number of these have assumed that the delay of

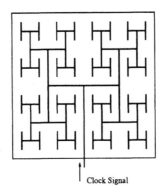

Clock Signal

Figure 5.12   H-tree clock routing

a source-sink path in the tree is proportional to the length of the path, i.e., that a zero *length skew* implies a zero delay skew. Under this assumption, the *H-tree* [105] provides a zero-skew clock routing tree, since the path length from the root to every sink node is the same (Fig. 5.12). The idea of the H-tree is generalized in [106] to be applicable to nonuniform distributions of sink nodes. The algorithm, called the Method of Means and Medians (MMM), constructs a clock tree hierarchically by recursively partitioning the set of sink nodes into two equal subsets and connecting the centers of the two subsets. It is shown that the maximum path length skew is bounded by $O(1/\sqrt{(n)})$ for n sink nodes in the unit square.

Another algorithm uses bottom-up tree construction based on recursive geometric matching to achieve zero length skew [107], while using less wire length than the MMM algorithm. Given a set S of $2n$ points on a plane, a *geometric matching* is a set of n nonintersecting edges whose endpoints are in S [108]. An optimal geometric matching is one in which the total wire length of the n segments is minimized. The clock routing algorithm of [107] uses heuristics to find a good geometric matching of the sink nodes, and joins the sink nodes with wire segments. For each wire segment, a point is found on the segment which is equidistant from the two sink nodes joined by the segment (Fig. 5.13). A geometric matching is then recursively computed on these points, to construct a tree with nearly perfect length balancing.

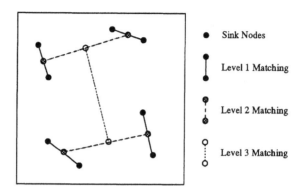

Figure 5.13 Recursive geometric matching

Zero-Skew-Algorithm$(S)\{$ /* S = set of sink nodes */
    for all $s \in S$, $Z = Z \cup s$ /* initialize zero-skew subtrees */
    while$(|Z| > 1)\{$
        pair up subtrees in Z;
        for each pair of subtrees (z_i, z_j):
            zero-skew-merge(z_i, z_j);
    $\}$
$\}$

Figure 5.14   Zero skew algorithm

5.6.1   Zero delay skew clock routing

The assumption that zero length skew implies zero delay skew is not valid in practice, due to the effects of wire resistance. The delay of a path is a function not only of the path length but also of the loading capacitances of branches attaches to the path. An algorithm for eliminating *delay skew* using the Elmore delay model is presented in [109]. The zero-skew algorithm is a recursive bottom-up procedure. The outline of the basic algorithm is illustrated in Fig. 5.14.

The key step in the algorithm is the *zero-skew merging* of two zero-skew subtrees to obtain another zero-skew tree. Each zero-skew subtree (ZST) z_i is characterized by a delay value t_i which is the delay from the root of the subtree to any leaf node (the delay is the same to any sink node in the ZST, by definition), and a capacitance value C_i which is the total capacitance of all nodes in

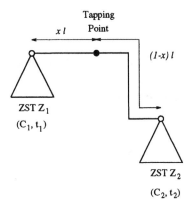

Tapping
Point

$x\,l$

$(1\text{-}x)\,l$

ZST Z_1

(C_1, t_1)

ZST Z_2

(C_2, t_2)

Figure 5.15   Zero-skew merging of two ZSTs

the ZST. Initially, every sink node forms a trivial zero-skew subtree, with $t_i = 0$ and $C_i = C_{sink}$. The root nodes of a ZST can be connected by a wire segment to obtain a new ZST, whose root lies at some point on the segment. To see how this can be done, consider two ZSTs z_1 and z_2, as shown in Fig. 5.15, with delays t_1 and t_2 and capacitances C_1 and C_2. Suppose the roots of the ZSTs are connected together by a wire of length l, with the tapping point (root of the new ZST) at a distance of xl from the root of z_1, where $0 < x < 1$.

For the merged tree to be a ZST, the delays from the tapping point to leaf nodes of both subtrees should be equal. Using the Elmore delay model and a lumped-π model for the wire segments, this can be written as

$$rxl(cxl/2 + C_1) + t_1 = r(1 - x)l(c(1 - x)l/2 + C_2) + t_2 \qquad (5.4)$$

where r and c are the resistance and capacitance per unit length of the wire.

Eq. 5.4 can be solved for x, the location of the tapping point:

$$x = \frac{(t_2 - t_1) + rl(C_2 + cl/2)}{rl(cl + C_1 + C_2)} \qquad (5.5)$$

If x lies between 0 and 1, the zero-skew merge is straightforward. A minimum length wire segment can be used to connect z_1 and z_2, and the location of the root of the new ZST can be computed using the value of x. However, if the values of t_1 and t_2 are widely different, x may be negative, or greater than 1. If $x < 0$, t_1 must be greater than t_2, so the location of the tapping point must be at the root of z_1. A wire of length l' must be used to connect z_1 to z_2, such

```
ZSTM(S){                                        /* S = {T₁, T₂, ..., Tₙ} */
    if |S| = 1 return(T₁);
    else if |S| = 2 return(zero-skew-merge(T₁, T₂));
    else {
        partition S into S₁ and S₂;
        Z₁ = ZSTM(S₁);
        Z₂ = ZSTM(S₂);
        return(zero-skew-merge(Z₁, Z₂));
    }
}
```

Figure 5.16   Zero skew segment tree method

that

$$t_1 = t_2 + rl'(C_2 + \frac{cl'}{2}) \tag{5.6}$$

This equation can be solved for l':

$$l' = \frac{\sqrt{(rC_2)^2 + 2rc(t_1 - t_2)} - rC_2}{rc} \tag{5.7}$$

If $x > 1$, the tapping point should be at the root of z_2, and the value of l' can be computed using an equation similar to Eq. 5.7.

The zero skew of the final tree is independent of any heuristics used for the pairing step in Fig. 5.14. Thus, these steps can be performed with a view to reducing total wire length or improving wireability. This fact is taken advantage of in [110] to generate zero-skew trees with shorter total wire length. Their algorithm, called the *zero skew segment tree method* (ZSTM), combines top-down partitioning of the set of sink nodes with the bottom-up zero-skew merging technique of [109]. The approach is outlined in Fig. 5.16.

The key observation in the ZSTM algorithm is that in the original zero-skew algorithm, the wire length increases whenever there is a large imbalance between the delays of two subtrees being merged, as evidenced by Eq. 5.7. To minimize the occurrence of such imbalances, the ZSTM algorithm recursively partitions the set S of sink nodes into two *balanced* subsets, such that the subtrees constructed over these two subsets are likely to have similar delays. The partition is chosen such that it can be represented by a simple curve separating the two subsets of points, and the total capacitances of the two subsets do not differ by more than $\max_{s \in S} C_s$, where C_s is the capacitance of sink s. Furthermore, among all partitions satisfying these criteria, the partition which minimizes the sum of the *diameters* of the two subsets of points is chosen. The diameter of a set of points S is defined as $\max_{s_i, s_j \in S} dist(s_i, s_j)$. The diameter of a set of points is a good indicator of the total wire length of a tree connecting the points. By minimizing the delay disparity between merged trees, the ZSTM algorithm is able to achieve a 10% reduction in wire length as compared to the algorithm of [109], while retaining the zero-skew property.

5.6.2   Reliable minimum-skew routing

The zero-skew algorithms described above eliminate skew by adjusting wire *lengths* to equalize source-sink path delays. However, the trees generated by such approaches are very sensitive to wire width variations which frequently occur in manufacturing processes. The *sensitivity* of sink delays to changes in wire widths is maximum for wire segments close to the clock driver node. Thus, "zero-skew" clock trees with small changes in wire width near the root may actually have large skews in practice. A different approach for *reliable skew reduction* is taken in [111], where the wire *widths* are adjusted to minimize delay, skew and skew sensitivity to process variations.

The procedure is based on an analysis of the sensitivities of Elmore delays to segment resistances and capacitances in a RC tree. Consider a pair of nodes i and j in a RC tree, such that i lies in T_j, the subtree rooted at j. The nominal delay t_i from the source to node i depends on the resistance and capacitance of branch j as follows:

$$\frac{\partial t_i}{\partial R_j} = C'_j \tag{5.8}$$

$$\frac{\partial t_i}{\partial C_j} = R_{c_{ij}} \tag{5.9}$$

where C'_j is the downstream capacitance at node j, i.e., the total capacitance of T_j, and $R_{c_{ij}}$ is the resistance common to the paths from the root to nodes i and j. If i is not in T_j, $\frac{\partial D_i}{\partial R_j}$ is zero, whereas the second expression remains unchanged. These expressions can be combined with equations for the resistance and capacitance as functions of wire width, to obtain node delay sensitivities with respect to branch widths. The delay sensitivities of all nodes with respect to all branches in the tree can be computed by path tracing, and they can be updated efficiently whenever any segment width is changed.

The skew minimization algorithm starts with an initial clock tree constructed using any method, and improves the tree in three stages:

1. **Delay reduction** In this phase, the delay sensitivities are used to identify wires which reduce the average delays when widened. An iterative improvement algorithm is used to widen selected wires in preset increments, until a target average delay specification is met.

2. **Desensitization** In this phase, wire widths are increased in bottom-up traversals of the tree, until the delay sensitivities are less than some specified value.

3. **Skew minimization** The tree at this point will have non-zero skews in general. To minimize skews, the algorithm iteratively identifies the best wire to widen, in terms of reducing the total deviation of all skews from the average skew value.

The insensitivity of clock trees generated using this approach to process variations was tested using Monte Carlo simulations. The worst case skews were found to lie between two and three times the nominal skew values.

5.6.3   Buffered clock trees

The large delays in system clock trees can be reduced by introducing buffers at certain nodes in the clock tree, to obtain a *buffered* or *multi-stage* clock tree (Fig. 5.17). The buffers increase current driving capability, and help to reduce delay by effectively isolating the path from the root to the buffer from the capacitance downstream of the buffer node.

The inclusion of buffer nodes modifies the computation of Elmore delays only slightly [109]. A buffer is characterized by an output impedance r_b, an input capacitance c_b and an intrinsic delay d_b. Thus, each branch i in the buffered RC tree must now include a delay value d_i, which is zero unless the branch is a buffer. The capacitance looking in to a branch i is c_b if the branch is a buffer; otherwise, it is equal to $c_i + C_i$, where c_i is the branch capacitance and C_i is the capacitance of the subtree rooted at node i. The delay at a node j, in terms of the delay at its parent i, is now given by

$$t_j = t_i + r_j C_j + d_j$$

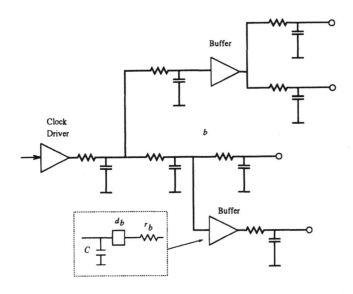

Figure 5.17   Buffered RC clock tree

The zero-skew algorithm of Fig. 5.14 is easily extended to the case of buffered clock trees [109]. The problem of buffer placement to minimize Elmore delays is discussed in [112]. In [113], algorithms are presented for even distribution of buffers over the routing area, so as to minimize routing congestion. The algorithms provide a tradeoff between delay, clock skew and total wire length.

The clock routing schemes discussed in this section are appropriate for VLSI, as they use only RC delay models. An interesting and challenging topic for future research is to extend these ideas to be accurate for MCM situations by considering transmission line effects.

5.7   SUMMARY

The problem of multilayer routing in MCMs can be decomposed into two sub-problems: two-dimensional routing and layer assignment. This chapter discussed the first subproblem, which is sometimes referred to as global routing, planar routing or tree construction. Performance-driven tree construction algorithms using RC and RLC delay models were described. Clock skew minimization under RC delay models was also discussed.

6

LAYER ASSIGNMENT APPROACHES

6.1 INTRODUCTION

Layer assignment is the process of determining which nets are to be routed on each layer of a multilayer routing substrate. There are several variants of this problem, corresponding to different physical characteristics of the routing environments in VLSI, printed circuit boards, and multichip modules.

The problem of layer assignment for VLSI routing has been studied extensively [90, 114, 115]. In most of this work, attention has been focused on minimizing the number of vias required to complete the routing in two layers, since the number of vias can increase the area and cost of a chip and reduce its manufacturability and reliability. For the two-layer situation, optimal polynomial-time algorithms have been proposed for minimizing the number of vias, based on the property that a max-cut for a planar graph can be found in polynomial time [116]. This result is true for the *constrained via minimization* approach on two layers, where two-dimensional route shapes for all the nets have been decided prior to layer assignment. The extension to three layers has been shown to be NP-complete [115]. The *unconstrained* or *topological* via minimization problem, in which rectilinear route shapes are decided after layer assignment, has also been shown to be NP-complete [114].

There is much less literature dealing with the layer assignment problem in the PCB environment, probably due to the fact that PCBs may have many more layers than VLSI, making the problem more complex and less interesting from a theoretical viewpoint. An early report on multilayer wiring investigated several innovative approaches for "unconstrained" layer assignment, but the experimental results were unimpressive: none of the sophisticated heuristics performed significantly better than very simple or even random layer assignment strategies [117]. Subsequent literature on the PCB wiring problem has focused on a particular wiring approach proposed by So [118], in which the overall problem is decomposed into a number of subproblems, including layer assignment, via column assignment and "single-row routing." These individual subproblems have also proven to be intractable [119]. A number of heuristics have been proposed for the single-row routing problem [120].

Layer assignment in ceramic and laminated MCM substrates is closely related to the PCB layer assignment problem, whereas the MCM-D substrate shares common features with VLSI. This chapter presents a model for the MCM-C layering environment, and describes a constrained layer assignment algorithm based on this model. A number of theoretical results are proved for the layer assignment problem, and experimental results on a number of randomly generated examples are presented, which demonstrate the effectiveness of the proposed algorithm.

For the higher density MCM-D environment, where a grid-based approach becomes less efficient, unconstrained layer assignment becomes a more attractive alternative. Section 6.3 presents an unconstrained layer assignment algorithm based on a new max-cut graph-partitioning heuristic and a congestion-based net interference graph generation approach, which improves significantly upon the results of [117].

6.2   LAYER ASSIGNMENT FOR CERAMIC MCMS

6.2.1   Previous work

Layer assignment for the MCM environment has received some attention lately, after the publication of one of the first papers dealing with MCM physical design [24]. The paper considered a layer assignment problem in multilayer ceramic substrates used in IBM's Thermal Conduction Modules (TCMs). Since the publication of [24], a number of papers on MCM routing have appeared, many of them using routing environment models which are based on the model proposed therein, or are extensions of it [22].

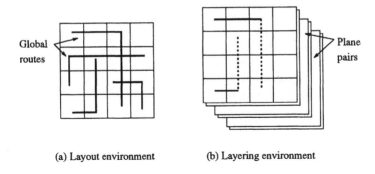

(a) Layout environment　　　　(b) Layering environment

Figure 6.1　Layout and layering environments

The model for the layer assignment problem in [24] consists of a *layout environment* and a *layering environment* (Fig. 6.1). The layout environment consists of a global routing grid, with two-dimensional global routes for all the nets specified on the grid. The maximum number of global routes crossing a vertical border of a tile in the grid is denoted by h_{max}, and the maximum number of routes crossing a horizontal border is denoted by v_{max}. The *density* of the problem, d_{max}, is defined as $\max\{h_{max}, v_{max}\}$.

The layering environment consists of several *x-y plane pairs*. Each plane pair contains one layer reserved for *x*-direction wiring and one layer reserved for *y*-direction wiring. Connections between layers of the same plane pair are established by *secondary vias*, whereas connections between plane pairs require *primary vias*, which are typically larger than secondary vias. Each plane pair is also tiled exactly like the layout environment, with the following *capacity constraint*: no more than ω wires may cross any tile boundary on a given plane pair.

The objective of the layer assignment problem is to assign every global route in the layout environment to a plane pair, such that the capacity constraint is satisfied on all plane pairs, and the number of plane pairs used is minimized.

A number of theoretical results about the problem are proved in [24]. The first establishes a lower bound on the number of plane pairs, L, required for a given problem:

$$L \geq \left\lceil \frac{d_{max}}{\omega} \right\rceil \tag{6.1}$$

The second result establishes the existence of instances of the problem which require as many as $3/2$ times this lower bound. The layer assignment problem is proved to be NP-complete even when all nets are two-terminal nets. Approximation algorithms for the two-terminal and k-terminal layer assignment problems (LA2 and LA-k) are presented, which establish upper bounds of $2 \left\lceil \frac{h_{max} + v_{max}}{\omega} \right\rceil$ and $(2k - 2) \left\lceil \frac{h_{max} + v_{max}}{\omega} \right\rceil$ on the number of plane pairs for the two problems, respectively.

One important factor which is ignored in this formulation is the maximum number of vias which can be allowed in any given tile. If a particular tile has a large number of primary vias (i.e, the tile is a terminal tile for a large number of nets), the number of secondary vias available in that tile becomes restricted, and a layer assignment which satisfies the capacity constraint may not be actually routable.

6.2.2   Multilayer model for the MCM-C routing environment

In a typical ceramic multichip module, all the chips are bonded to the top layer (chip layer) of a multilayer substrate. Below the chip layer, there is a stack of wiring layers in which all the chip-to-chip nets must be interconnected, connections to module I/O pins (located on the bottom layer) must be established, and power and ground connections to all the chips must be provided. The number of wiring layers may be as high as 60 or more, as reported in [4]. The problem of multilayer routing has been studied extensively for the printed circuit board (PCB) and integrated circuit environments. The standard models used for these two environments are described as follows [121]:

1. *Single Active Layer (SAL) model.* This model, used for the VLSI routing environment, consists of two to four wiring layers, with all net terminals present on the top or "active" layer. Paths connecting a pair of terminals must begin and end on the active layer, but may change layers any number of times, by contact cuts or "vias." Paths of different nets cannot cross on any layer.

2. *Printed Circuit Board (k-PCB) model.* In this model, there are k wiring layers. All the holes are "through holes," i.e., a terminal of a net is available on all k layers at the same $x - y$ location, so that a path may begin and end on any layer. Each layer contains a grid of $x - y$ locations. Paths may

change layers through vias at $x - y$ locations which have not been used as vias by any other paths. Paths of different nets cannot cross on any layer. Usually, individual layers are reserved for wiring in only one of two orthogonal directions [122].

To capture some of the features unique to the MCM-C environment, we introduce a new $k - MCM$ model. In this model, there are k wiring layers, including two active layers - the top and the bottom - corresponding to the chip pads on the top layer of an MCM substrate and the module I/O pins on the bottom layer. As in other models, paths of different nets cannot cross on any layer, and each layer is a grid of $x - y$ locations. The model does not assume that holes are drilled through all the k layers. A path on layer m may change layers at any $x - y$ location which is not used as a via between layers l and n, $l \leq m \leq n$, by any other net. This allows the same $x - y$ location to be used as a via by different nets on different layers, thus reflecting the availability of segmented vias in the MCM fabrication technology (cf. Fig. 4.3(b)).

6.2.3 Layer assignment in the k-*MCM* model

This section presents a formulation of the layer assignment problem for the $k - MCM$ model. The formulation allows bidirectional wiring on all the layers and segmented via usage. To distinguish this problem from other layer assignment problems discussed in the literature, we will refer to this problem as the *Detailed Layer Assignment (DLA)* problem, since it is formulated on the detailed wiring grid. The input to the DLA problem describes a grid of points called a hole array and a set of wires or routes. Each wire consists of a number of horizontal and vertical lines, which interconnect a set of holes corresponding to the terminals of some net in the interchip netlist. The lines are drawn in tracks that pass between the rows and columns of the hole array. Short legs join the endpoints of lines to appropriate holes (Fig. 6.2).

The output of DLA is a detailed multilayer routing for the entire set of connections, obtained by assigning every segment of each wire to a unique layer. The final layer assignment must satisfy the following constraints:

(C1) On any layer, at most one line can occupy a track at any point,

(C2) Intersecting lines of different nets must be assigned to different layers, and

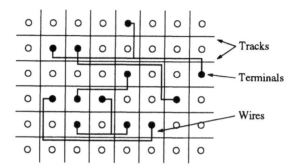

Figure 6.2   Input to the DLA problem

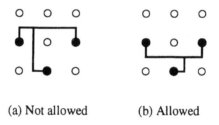

(a) Not allowed   (b) Allowed

Figure 6.3   Restriction on input routes

(C3) On any layer, a hole can be used as a via by at most one net.

Constraints on timing and crosstalk can also be included, as discussed in Section 6.2.7. We first consider the case of a single track between adjacent rows or columns of holes. To simplify the formalization of the DLA problem, the following mild restriction is imposed on the routes: a connection between a route and a hole can be made only if the route is in a track immediately below or to the right of the hole. Thus, the route in Fig. 6.3(a) is not allowed, but the equivalent route in Fig. 6.3(b) is allowed. It is easy to see that this is not a severe restriction, since a route of type (a) can be easily modified with only minor changes in its length and shape into a route of type (b).

Every hole location in the fixed hole array is indexed by an ordered pair of positive integers (i, j), where i and j are the row and column numbers of the location. The route for each net is decomposed into a connected set of wire segments (Fig. 6.4). Each wire segment is either a horizontal link joining the two adjacent hole locations in the row above it, or a vertical link joining the

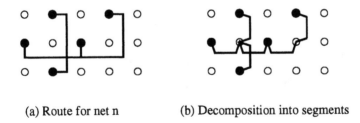

(a) Route for net n   (b) Decomposition into segments

Figure 6.4   Decomposition of routes into wire segments

two adjacent hole locations in the column to the left of it. Every wire segment belonging to a net n is labeled n.

Given a set of decomposed routes, a labeled directed multigraph $D(V, E)$, called the *Segment-Location Digraph (SLD)*, is constructed in the following way: every location (i, j) in the fixed hole array is represented by a unique vertex $V_{ij} \in V$. Every horizontal segment connecting two locations (i, j) and $(i, j + 1)$ is represented by a labeled directed edge from vertex V_{ij} to $V_{i,j+1}$. Every vertical segment connecting two locations (i, j) and $(i+1, j)$ is represented by a labeled directed edge from vertex V_{ij} to $V_{i+1,j}$. The label of the edge is the same as the label of the segment. If a location (i, j) is a top-layer terminal of some net n, vertex V_{ij} is labeled T_n. If (i, j) is a bottom-layer terminal of net n, V_{ij} is labeled B_n. Any vertex can have at most one T-label and one B-label (a vertex may have both labels simultaneously). Figure 6.5 shows a set of routes, and the SLD derived from it. The set of directed edges of label n entering or leaving a vertex V_{ij} is called a *segment group of net n adjacent to V_{ij}*, denoted by $S_n(i, j)$. An *edge coloring* of the SLD is a function, $f : E \to \{1, 2, \ldots\}$. An edge coloring induces a set of intervals at every vertex in V, one interval for every segment group adjacent to the vertex. The intervals are defined as follows: at any vertex V_{ij} labeled T_n, or $T_n B_m, m \neq n$, group $S_n(i, j)$ has an associated interval

$$[1, \max_{a \in S_n(i,j)} f(a)] \tag{6.2}$$

If V_{ij} is labeled T_m or $T_m B_l, m, l \neq n$, or is not labeled, the interval associated with $S_n(i, j)$ is empty if $f(a) = f(b)$ for all $a, b \in S_n(i, j)$; otherwise, it is given by

$$[\min_{a \in S_n(i,j)} f(a), \max_{a \in S_n(i,j)} f(a)] \tag{6.3}$$

If V_{ij} is labeled B_n or $T_m B_n, m \neq n$, the associated interval is

$$[\min_{a \in S_n(i,j)} f(a), \infty) \tag{6.4}$$

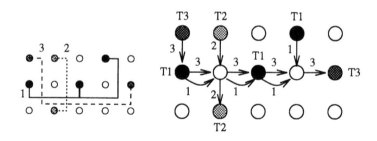

(a) Given routes for 3 nets   (b) SLD derived from the routes

Figure 6.5 Construction of the SLD

and if it is labeled $T_n B_n$, the associated interval is

$$[1, \infty) \tag{6.5}$$

A set of intervals is *overlapping* if some pair of intervals associated with groups of different nets overlap; otherwise, the set is disjoint. Figure 6.6 illustrates the intuition behind these definitions - the basic idea is that disjoint intervals correspond to conflict-free segmented via usage. If all segments in a group are placed on the same layer, they can be routed without touching the hole that the group is adjacent to (unless the hole is a terminal of the net). This is why the interval at a nonterminal vertex is defined to be empty when all edges in a group are colored the same. If a location (i, j) is a top *and* bottom layer terminal of the same net, it cannot be used as a via by any other net on any layer.

The goal of the DLA problem is to find an edge coloring, $f : E \rightarrow \{1, 2, \ldots\}$, of the SLD, using the minimum number of colors, such that

(D1) At any vertex, no two outgoing edges with different labels have the same color.

(D2) At every vertex, the set of intervals induced by the coloring is disjoint.

From an edge coloring f, a layer assignment can be determined in a straight-forward way: segment u is assigned to layer $f(u)$. If f satisfies (D1) and (D2), the corresponding layer assignment can be shown to satisfy the constraints (C1)-(C3) mentioned earlier. Thus, the edge coloring f gives us a legal layer assignment. The general detailed layer assignment problem with terminals on

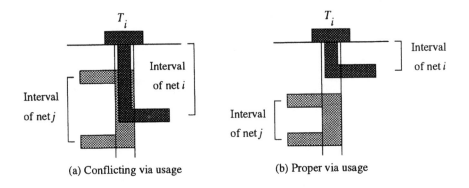

Figure 6.6 Intuition behind interval definitions

top and bottom layers, referred to as 2DLA, has been shown to be NP-complete in [123].

Theorem 6.1 *2DLA is NP-complete, even if $K = 3$ and edges with at most two different labels are incident at any vertex.*

6.2.4   Heuristics for DLA

In this section, we consider two top-to-bottom layer-by-layer heuristics to solve the DLA problem. These approaches begin with the given DLA instance, and find a 0-1 edge coloring to maximize some objective. The edges colored 1 are assigned to the top layer, and are deleted from the DLA instance. The vertex labeling is then updated as described below, to obtain a new DLA instance for the second layer. The procedure is then repeated for successive layers, until there are no edges remaining in the graph. The 0-1 coloring solutions must satisfy the following constraints:

(E1) At any vertex, two outgoing edges with different labels cannot both be colored 1. At any vertex labeled $T_i B_j, j \neq i$, no edge labeled j can be colored 1.

(E2) At any unlabeled vertex, edges of at most one label can be dichromatic. (A group of edges is *dichromatic* if at least one edge in the group is colored

1 and at least one edge is colored 0. If a group of edges labeled i and incident at V is dichromatic, we say that net i is dichromatic at V.)

(E3) At any vertex labeled T_i or $T_i B_j$, (j may be equal to i), only net i can be dichromatic.

(E4) At any vertex V labeled B_i, if an edge labeled i is colored 1, no net $k, k \neq i$, can be dichromatic at V.

After performing the coloring, the edges colored 1 are assigned to the current layer and deleted, to obtain a problem with fewer edges for the next layer. To account for the via usage caused by splitting a route for a net into partial routes on different layers, the vertex labels for the DLA problem on the next layer must be updated as follows:

(U1) If a net n is dichromatic at an unlabeled vertex V_{ij}, V_{ij} is labeled T_n.

(U2) Vertices labeled $T_n B_n$ are left unchanged. Vertices labeled T_n or $T_n B_m, m \neq n$, are left unchanged if any edge labeled n is colored 0 at the vertex; otherwise, the label is changed to B_m.

(U3) A vertex labeled B_n is changed to $T_n B_n$ if any edge labeled n is colored 1 at the vertex; otherwise, the label is left unchanged.

The constraints (E1)-(E4) and the vertex labeling update rules (U1)-(U3) ensure that the resulting layer assignment is legal.

We consider two different optimization objectives for the 0-1 coloring problem on each layer:

Maximum routable subset of edges

This approach attempts to minimize the number of layers by maximizing the total length of wires routed on each layer, starting with the first layer. Although the problem of maximizing the wire length on the current layer appears to be simpler than the original DLA problem, it is shown in [123] that it is also NP-complete.

Consider the following decision problem, which we shall refer to as the Maximum Routable Subset (MRS) problem:

Maximum Routable Subset

Instance: A labeled directed multigraph $D(V, E)$ derived from an array of hole locations and a set of routes, and an integer $K \leq |E|$.

Question: Is there a coloring $f : E \rightarrow \{0, 1\}$, which satisfies (E1) through (E4), such that at least K edges are colored 1?

Theorem 6.2 *MRS is NP-complete, even if edges of at most three different labels are incident at any vertex and all terminals are top layer terminals.*

Maximum outdegree reduction

The labeled outdegree $d_o(v)$ of any vertex $v \in V$ is defined to be the number of outgoing edges with different labels at the vertex. The maximum labeled outdegree Δ_o gives a simple lower bound on the number of layers required for solving an instance of DLA. Thus, a good approach would be to reduce Δ_o by at least 1 in the minimum number of layers (ideally, in one layer) to obtain a new DLA instance with maximum labeled outdegree at most $\Delta_o - 1$. Repeating this procedure until $\Delta_o = 0$ will tend to produce a layer assignment using the minimum number of layers. We do not know if the problem of deciding whether Δ_o can be reduced by at least 1 in at most K layers (where $K < |S| = |\{v|v \in V, d_o(v) = \Delta_o\}|$) is polynomial-solvable. Clearly, if $K = |S|$, then the answer is always "yes." However, this trivial upper bound is "tight," because there exist classes of instances of DLA with bounded Δ_o which achieve this bound. Figure 6.7 shows an example of such a class of instances. The hole array is $n \times n$, and there are $(n/3)$ vertices with $d_o(v) = \Delta_o = 3$. There are n three-terminal T-shaped nets, and all holes which are not terminals of these nets are unavailable for use as vias, as they are top and bottom layer terminals of two-terminal nets comprised solely of one top layer terminal and one bottom layer terminal.

This example proves two facts for arbitrary instances of DLA:

Fact (1) Let L be the number of vertices with $d_o(v) \geq 1$. For the maximum routable subset heuristic, no algorithm can guarantee to color at least rL edges for any constant $0 < r \leq 1$. In the example, $L = O(n^2)$, whereas the maximum number of edges that can be colored 1 is $O(n)$, so the ratio r can be made arbitrarily small by increasing n.

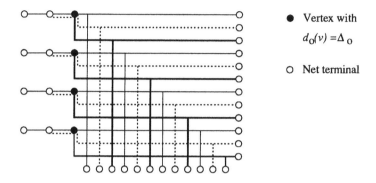

Figure 6.7   DLA instance with $\Delta_o = 3$

Fact (2) For the maximum outdegree reduction heuristic, no algorithm can guarantee to reduce Δ_o by at least 1 in $\leq c(\Delta_o)$ layers, for any constant $c(\Delta_o)$. In the example, at least $(n/3)$ layers are required to reduce Δ_o from 3 to 2, since no two routes can be simultaneously colored 1. Thus the number of layers required can be made arbitrarily large by increasing n.

Although the conclusions of Theorems 5.1 and 5.2 and Facts (1) and (2) appear discouraging, it should be noted that their proofs are based on particularly difficult instances of DLA. For example, the proofs of Facts (1) and (2) are based on instances with a very high density (asymptotically 100%) of top and bottom layer terminals. In the next section, we describe some experimental results using the maximum outdegree reduction approach, which indicate that very good solutions can be obtained for practical situations.

6.2.5   A heuristic "cluster-coloring" algorithm

In this section, we describe a heuristic solution to the maximum outdegree reduction problem. We begin by defining clusters of a net (to simplify the presentation, an edge will be called "colored" if it is colored 1, and "uncolored" if it is colored 0).

Definition of clusters

In general, edges of a particular net cannot all be colored independently to obtain a valid coloring (i.e., a coloring that satisfies (E1)-(E4)). For example, consider two edges u and v of net i, incident at a vertex labeled T_j, $j \neq i$. Both u and v must be colored the same (either both 0 or both 1), because net i is not allowed to be dichromatic at T_j. Based on this observation, we define a *cluster* of net i, denoted C_i, to be a maximal connected set of edges of label i which must all be colored the same to obtain a valid coloring. A *potential via* of net i is defined as a vertex at which net i is allowed to be dichromatic. Thus, a cluster of net i is a subtree of net i, such that each leaf of the subtree is a potential via, and no internal vertex of the subtree is a potential via. Two clusters of a net are defined to be *connected* if they share a leaf vertex.

When no edges of any net are colored, every net i is partitioned into a unique set of clusters, determined only by the vertex labeling. This unique partitioning is called the *primary partitioning*, and the clusters are called the *primary clusters*. If some of the edges are colored, this may affect the partitioning. For example, if an edge labeled i incident at a vertex labeled B_i is colored, no other net can be dichromatic at that vertex. So any edges labeled $j, j \neq i$ incident at the vertex will be forced to belong to the same cluster C_j. Thus, any coloring of the edges induces a partitioning of the nets into induced clusters. The following lemma gives the relation between the clusters induced by two different colorings:

Lemma 6.1 *Given two valid colorings f and g, such that (a) every colored edge in f is also colored in g, and (b) for any i, every colored edge labeled i in g is either colored in f or is not connected to any colored edge labeled i in f; then every cluster induced by g is the union of one or more connected clusters induced by f.*

Corollary 6.1 *Given any valid coloring f, each cluster induced by f is the union of one or more connected primary clusters.*

The proof of the lemma follows from the observation that a valid coloring of a set of edges can only reduce the set of potential vias available for other nets. We define a cluster to be colorable if every edge in the cluster can be colored without violating (E1). Finally, we define a *Maximal Connected Colorable Set* (MCCS) of net i as a maximal connected set of clusters of net i, such that every cluster in the set is colorable.

Cluster-coloring heuristic

The definition of clusters motivates a sequential heuristic algorithm for DLA, based on the maximum outdegree reduction approach. The algorithm begins

```
0.     layer = 1;
1.     generate primary clusters();
2.     for k = 1 to Δ_o  V_k := {v ∈ V|d_o(v) = k};
3.     for k = Δ_o to 1 {
4.         for each v ∈ V_k {
5.             c = find colorable cluster(v);
6.             color MCCS(c);
7.             update clusters();
8.         }
9.     }
10.    delete colored lines();
11.    update vertex labels();
12.    if(not all colored) layer := layer+1; goto 1;
13.    stop;
```

Figure 6.8   Cluster coloring algorithm

with the primary clusters on the top layer. For each vertex v with labeled outdegree $d_o(v) = \Delta_o$, the algorithm searches for a colorable cluster containing an outgoing edge at v. If a colorable cluster c is found, every cluster in the MCCS containing c is colored. Then, by Lemma 6.1 and the definition of a MCCS, each cluster induced by this coloring will be the union of one or more existing clusters. The appropriate clusters are merged to form new clusters, and the process is repeated, visiting vertices in decreasing order of d_o (Fig. 6.8). By the definition of clusters, the procedure inductively generates a coloring which satisfies (E2)-(E4) also. After deleting all colored edges and updating the vertex labeling, the entire procedure is repeated for succeeding layers, until no uncolored edges remain.

If more than one colorable cluster is found at a vertex, the algorithm favors a cluster which belongs to a short MCCS containing the maximum number of top layer terminals of the net. This tends to reduce the number of uncolorable clusters (since a large MCCS interferes with a greater number of clusters than a smaller MCCS, on the average), thus increasing the probability that the maximum outdegree can be reduced in a single layer. Favoring clusters with many top layer terminals tends to reduce the number of vias by reducing the distance that vias must travel.

6.2.6   Experimental results and extensions

Example problems with random netlists

The performance of the layer assignment algorithm was measured on three sets of random netlists, which were generated to resemble actual circuit netlists, with a realistic distribution of multiterminal nets with 2-6 terminals per net.

Table 6.1 shows the statistics of the 12 examples. In the first set of 5 examples (ex1 - ex5), the percentage of active pins ($x - y$ locations that were top or bottom layer pins) is about 30%. In the second set of examples (ex6 - ex10), the active pin percentage (APP) is very high, typically $\geq$ 85%, to simulate a very densely packed MCM with VLSI chips using flip-chip bonding. The third set (MCM213, MCM848) [124] consists of two large examples with 210 and 842 nets, respectively, with all top layer terminals and APPs of 43% and 49%.

The two-dimensional input routes for the nets for all the examples were generated by a simple router which attempted to minimize the maximum routing congestion. The routing for the largest example was generated in about 10 minutes on a HP 9000/845 Superworkstation.

DLA results

Number of layers

Table 6.2 shows the lower bound (Δ_o) on the number of layers and the number of layers actually used by the algorithm. In most cases, the number of layers could be reduced by a simple maze router applied to the small number of partial nets remaining on the final two layers, because the final layers are sparsely populated, with a small number of nets concentrated in the center region of the layout. In such a situation, the restrictions imposed in Section 6.2.3 may be removed to improve the routability, thereby allowing the layer count to be reduced. The third row in Table 6.2 gives the final number of layers after this post-processing.

For comparison, the lower bound on the number of layers using a unidirectional layer approach [24] is also shown in the last row of the table. The number of layers used by the DLA approach is always less than the *lower bound* on the number of unidirectional layers, for the first and third sets of examples. For the second set, the high APP makes the lower bound an unrealistic estimate

Table 6.1   Statistics of 12 randomly generated examples

Example	ex1	ex2	ex3	ex4	ex5
Rows×Cols	15 × 15	20 × 20	25 × 25	30 × 30	40 × 40
Nets	25	40	75	100	200
APP	31%	30%	33%	33%	35%

Example	ex6	ex7	ex8	ex9	ex10
Rows × Cols	10 × 10	15 × 15	20 × 20	25 × 25	30 × 30
Nets	30	70	120	200	275
APP	78%	86%	88%	88%	86%

Example	MCM213	MCM848
Rows × Cols	40 × 40	75 × 75
Nets	210	842
APP	43%	49%

of the number of layers required, since a unidirectional layer approach would be severely constrained by the shortage of vias. For these examples, a more realistic estimate of the number of layers required is also shown in the last row, which is based on the availability of vias on each layer and the average number of bends per net. Again, the DLA approach always uses fewer layers.

Number of vias

Table 6.3 shows the total number of primary and secondary via holes required in the substrate for the DLA-based routing as well as the estimated number of via holes for a unidirectional layer based approach. In the latter approach, every bend in a net requires a secondary via, and every terminal requires $(L/2)$ primary vias on the average, where L is the number of layers. The DLA approach uses significantly fewer vias.

Table 6.2 Number of MCM layers

Example	ex1	ex2	ex3	ex4	ex5
Lower Bound	4	5	6	7	9
Layers Used	4	6	7	8	11
Final Layers	4	5	6	8	10
Lower Bound (Uni.)	6	6	8	10	12

Example	ex6	ex7	ex8	ex9	ex10
Lower Bound	5	8	11	12	14
Layers Used	6	10	15	18	21
Final Layers	5	9	13	17	20
Lower Bound (Uni.)	8	10	12	16	18
Est. Layers (Uni.)	8	12	16	20	22

Example	MCM213	MCM848
Lower Bound	7	9
Layers Used	8	11
Final Layers	7	10
Lower Bound (Uni.)	10	14

Table 6.3 Total number of vias

Example	ex1	ex2	ex3	ex4	ex5
Vias (DLA)	210	373	1063	1967	4592
Est. Vias (Uni.)	425	723	1607	2791	6194

Example	ex6	ex7	ex8	ex9	ex10
Vias (DLA)	205	983	2688	5903	9910
Est. Vias (Uni.)	480	1706	4060	8063	12347

Example	MCM213	MCM848
Vias (DLA)	2766	16639
Est. Vias (Uni.)	5741	26743

Table 6.4 Run time

Example	ex1	ex2	ex3	ex4	ex5
Run time (sec)	1.0	3.3	10.1	24.5	104.7

Example	ex6	ex7	ex8	ex9	ex10
Run time (sec)	0.8	3.5	12.7	34.6	76.3

Example	MCM213	MCM848
Run time (sec)	58.6	899.8

Run time

The complexity of the DLA algorithm is $O(kmn)$, where k is the number of layers used, and m and n are the numbers of rows and columns in the detailed routing grid. Table 6.4 shows the run time for the 12 examples. The largest example took about 15 minutes on a HP 9000/845 Superworkstation.

6.2.7 Extensions

Incorporation of timing and crosstalk constraints

Since the DLA approach imposes only mild restrictions on the two-dimensional shapes of routes, a performance-driven router can be used to generate the input routes. The timing delays in the DLA-based MCM routing can be further reduced by taking net criticality into account while searching for a colorable cluster at a vertex. Clusters of more critical nets can be considered for coloring in preference to those of less critical nets, so that more critical nets will have a better chance of being routed closer to the top active layer.

Crosstalk constraints can be included as extra constraints to be considered while deciding whether a cluster is colorable. Given a set of nets that are mutually sensitive to crosstalk, coloring a cluster of any one of these nets will make all the clusters of the remaining nets uncolorable. This prevents interfering nets from being placed on the same layer.

Extension to multiple tracks

The DLA formulation can be extended to the case of p wiring tracks between adjacent $x - y$ locations, $p > 1$, by introducing $p - 1$ extra rows (columns) of *virtual locations* between adjacent rows (columns) of existing $x - y$ locations, and labeling all virtual locations as $T_x B_x$, so that they cannot be used as vias by any net. This approach increases the run time for a given problem by approximately p, since m and n each increase by a factor of p, and k decreases by a factor of approximately p.

6.3 LAYER ASSIGNMENT FOR HIGH-DENSITY MCMS

The DLA approach of the previous section is based on a routing grid. Hence, it is impractical for high-density MCMs, since the size of the grid becomes unmanageable. In this section, we present an alternate approach to layer assignment, which does not require a grid. The approach is applicable to both constrained and unconstrained layer assignments. Given a set of nets, with or without their two-dimensional global routes, the approach generates a *net interference graph (NIG)*. Each vertex in the NIG represents a net in the netlist, and the weight of an edge (i, j) joining two vertices represents an estimate of the likelihood of a nonplanar situation if the nets i and j are placed on the same layer, as well as the potential crosstalk between the two nets. The vertex set V of the NIG is then partitioned into k subsets, $V_1, \cdots, V_k$, using a new linear-time partitioning heuristic. Subset V_i represents a set of nets to be assigned to the ith layer. The value of k may be specified by the user. The partitioning algorithm attempts to maximize the following quantity:

$$\sum_{(i,j)\in E} w_{ij} \cdot \Delta(C_i, C_j) \tag{6.6}$$

where

$$\begin{aligned} \Delta(C_i, C_j) &= 0 \text{ if } C_i = C_j \\ &= 1 \text{ if } C_i \neq C_j \end{aligned}$$

where $w_{i,j}$ is the edge weight (potential interference) between two vertices (nets) i and j and C_i is a color (layer) assigned to vertex (net) i.

In some MCM substrates, adjacent signal layers may not be separated by ground planes. In such substrates, wires may interfere electrically even if they are placed on different layers. To further reduce crosstalk in such situations, a *layer permutation* algorithm is presented, which maximizes the layer separation between interfering nets, so as to reduce interlayer crosstalk. The objective of the layer permutation algorithm is to maximize

$$\sum_{(i,j)\in E} w_{i,j} \cdot |C_i - C_j|$$

Unlike the DLA approach, the majority of nets in this approach are constrained to be routed in a single layer. The formulation of the NIG and the partitioning ensure that the set of nets assigned to each layer forms a planar set, or a nearly planar set. If the unconstrained approach is used, layer assignment is followed by single-layer routing on each layer. Any nets which remain unrouted must then be completed by allowing layer changes with vias.

Figure 6.9 gives an overview of the layer assignment approach. Section 6.3.1 describes the construction of the NIG, and Section 6.3.2 describes the layer partitioning and permutation algorithms.

Input: A netlist of N nets
Output: An assignment $A : N \rightarrow L$
(Step 1) Construction of NIG, G.
(Step 2) Vertex partitioning and color permutation
(Step 3) Single-layer routing on each layer
(Step 4) 3-D routing for unrouted nets.

Figure 6.9 High-density MCM layer assignment algorithm

6.3.1 Construction of the net interference graph

Given a set of nets N, a graph $G(V, E)$ is constructed as follows. There is a one-to-one correspondence between vertices in V and nets in N, and an edge in E represents *potential interference* between corresponding nets in N. The graph G is called a *net interference graph (NIG)*. The purpose of the NIG is to model the interference between nets as pairwise interactions. The stronger the interference between a pair of nets, the less desirable it is to place both the nets on a single layer. There are two components of interference that we consider here: the first is *potential nonplanarity*, which estimates the likelihood of a nonplanar situation arising if both nets are placed on the same layer. In the case of constrained layer assignment, this component is simply 1 if the nets intersect, and 0 if they do not. For unconstrained layer assignment, where the global routes of the nets are not known in advance, the computation is more complicated, as described below.

The second interference component is *potential crosstalk*. This is an estimate of the electrical crosstalk that may occur between a pair of nets if they are placed on the same layer. The computation of both components is described below.

Potential nonplanarity

For the constrained layer assignment case, a nonplanar situation is caused by two nets if and only if their routes intersect. Given the global routes for all the nets, the interference graph can be found by a plane sweep in $O(n \log n + k)$ time, where k is the number of all intersecting pairs and n is the number of wire segments.

For the unconstrained situation, potential nonplanarity between two nets can only be estimated, since the global routes are not known at the time of layer assignment. Ideally, the nonplanarity measure used to generate the NIG should represent a necessary and sufficient condition for planarity: a set of nets should be planar if and only if the corresponding set of vertices in G is an independent set. This is not possible in general because interference is not a pairwise property, and can thus only be approximated using a graph model. For example, consider the three nets A, B and C in Fig. 6.10. Clearly, placing all three nets on the same layer results in an unroutable situation; however, any two of the nets can be routed on the same layer. Thus, it is not possible to model the interaction of these three nets as a pairwise interaction.

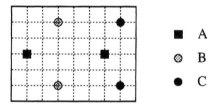

Figure 6.10   Three interfering nets

It is easy to come up with interference measures which satisfy the sufficient condition, i.e., if a set of vertices in G is independent, the corresponding set of nets is planar. For example, a simple measure which satisfies this property adds an edge between two vertices iff the bounding boxes of corresponding nets intersect.

A number of interference measures were compared in [117], and the conclusion of the author was that no single measure performed significantly better (in terms of the percentage of nets successfully routed after layer assignment) than the others. However, none of the measures considered the effect of *congestion*. If the bounding boxes of two nets intersect in a highly congested region, the routability will be more severely affected than if they intersect in a region with very few nets. Based on this observation, we propose a new interference measure.

A *bounding box (bbox)* of a net is defined to be the smallest rectangle which contains all the terminals of the net. A *super-bounding box (sbox)* of a pair of nets is the union of the bounding boxes of the two nets, and an *intersection box (ibox)* of a pair of nets is the intersection of the two bounding boxes. Let

$w(i, j)$ denote the weight on the edge (i, j) in the NIG. The weight is calculated as follows:

$$w(i, j) = \frac{(\rho(i) + \rho(j) + \eta(ibox)) * A(ibox)}{A(bbox(i)) \, A(bbox(j))}$$

where $\rho(i)$ is the number of nets whose bounding boxes intersect the bounding box of net i, $\eta(ibox)$ is the number of nets which have a terminal in *ibox*, and *A (box)* denotes the area of a box. The reason for dividing by the areas of the bounding boxes is that a larger area makes it easier to find nonintersecting routes for nets i and j, thus reducing the interference. It is easy to see that this measure satisfies the sufficient condition mentioned earlier.

The effectiveness of this interference measure was evaluated experimentally on a number of random netlists. For a given netlist, NIGs were generated based on four different interference measures. Then, layer assignment was performed using the max-cut k-color partitioning algorithm, followed by maze routing on each layer. The different interference measures were compared based on the percentage of nets successfully routed.

Table 6.5 shows the comparison between the following interference measures:

1. $w(u, v) = 1$ if the bounding boxes of u and v intersect.

2. $w(u, v) =$ area of *ibox* of u and v.

3. $w(u, v) =$ number of nets intersecting the *sbox* of u and v, divided by the area of the *sbox*.

4. $w(u, v) = \dfrac{(\rho(u) + \rho(v) + \eta(ibox)) * A(ibox)}{A(bbox(i)) \, A(bbox(j))}$.

The number of layers allowed for each example was obtained as follows: If too few layers are used, the routing problem on each layer will be unrealistically difficult. On the other hand, if too many layers are used, so that the completion rate is close to 100%, then it will be difficult to differentiate a poor interference measure from a good one. Therefore, the number of layers is adjusted for each example so that completion rate for measure (1) (the simplest measure) is about 70%.

The results show that the new interference measure (4) based on congestion performs better than the others in most of the cases.

Table 6.5 Comparison between different interference measures

Interf. Measure	(1)	(2)	(3)	(4)
Ex (#nets /# layers)	Routing Success (%)			
1 (100/6)	51	48	52	54
2 (60/6)	77	70	75	73
3 (38/4)	74	71	71	84
4 (40/4)	63	63	70	68
5 (50/4)	66	64	70	68
6 (75/5)	63	69	61	71
7 (36/3)	67	69	72	72
8 (47/5)	70	70	70	72
9 (50/4)	72	66	68	76
10 (62/4)	65	57	60	65
Average	66.65	64.69	66.97	70.27

Potential crosstalk

For the constrained layer assignment case, the maximum potential crosstalk between two nets is estimated as shown in Fig. 6.11. The shaded region corresponds to the set of crosstalk-critical regions induced by the given global routes of the two nets. All crosstalk-critical areas are extracted by performing two plane sweeps, one for parallel horizontal segments scanning the plane in the x-direction and another for parallel vertical segments scanning the plane in the y-direction. For example, in Fig. 6.11, $R_1 = (L_1, W_1)$ and $R_5 = (L_5, W_5)$ are extracted during the y-scan, and the remaining regions during the x-scan. The crosstalk is proportional to the maximum length for which two nets run in parallel, and is inversely proportional to the minimum separation between the parallel wires:

$$\chi = \sum_i (L_i / W_i) \tag{6.7}$$

For the unconstrained layer assignment approach, the maximum potential crosstalk is estimated as shown in Fig. 6.12. The crosstalk is proportional to the estimated length for which the two nets may run in parallel, and inversely proportional to the maximum estimated separation between the parallel wires, so as to avoid overly pessimistic estimates. Three cases may be identified, depending on the type of intersection between the net bounding boxes:

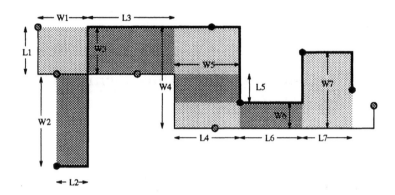

Figure 6.11   Crosstalk estimation for constrained layer assignment

Case 1: There is an intersection (sharing four points) between the bounding boxes (Fig. 6.12 (a)), χ = semiperimeter of *ibox* / $(W_1 + W_2)$;

Case 2: There is an intersection (sharing two or three points) between the bounding boxes (Fig. 6.12(b)), χ = semiperimeter of *ibox* / semiperimeter of *sbox*; and

Case 3: There is no intersection between the bounding boxes (Fig. 6.12 (c)), $\chi = L/W$

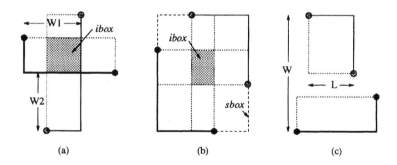

Figure 6.12   Crosstalk estimation for unconstrained layer assignment

The weight of edges in the NIG is computed as a linear combination, $\alpha\chi + (1 - \alpha)H$, where H denotes potential nonplanarity and χ denotes potential crosstalk. The parameter α reflects the relative importance of minimizing the number of layers and minimizing crosstalk.

6.3.2   Partitioning and color-permutation algorithms

Partitioning

The problem of maxcut k-color partitioning can be solved using the approximation algorithm of [125], which is described briefly in this section. The algorithm is built on the *binary decomposition tree* paradigm. The idea is to bipartition the graph into two subgraphs recursively, producing a maximum cut between the two subgraphs at each step (equivalently, minimizing interference inside each set). However, since finding a max-cut is NP-complete in general, an approximation algorithm is used, which runs in $O(|V| + |E|)$ sequential time yielding a *guaranteed* cut size of at least $(|E| + |E|/|V|)/2$. The subgraphs obtained are further bipartitioned until we get a total of k subgraphs. A subgraph is not partitioned further if it does not contain any edges. The algorithm generates a k-color partition guaranteeing that the total cost C, given by the total number of edges in the k induced subgraphs, is bounded as follows:

$$C \le \frac{|E|}{k} \left[\frac{|V| - 1}{|V|} \right]^{\log k} \qquad (6.8)$$

where $|E|$ is the total number of edges in the graph and k is the number of partitions. The tree generated in the course of the hierarchical partitioning procedure is called a *maxcut color tree*.

In the layer assignment formulation, the graph under consideration has weighted edges. For weighted graphs, the same algorithm can be used for k-partitioning, with similar performance guarantees: the number of edges $|E|$ in Eq. (6.8) is replaced by the sum of edge weights $\sum_{e \in E} w(e)$. The solution quality of this algorithm was tested experimentally and found to be competitive with simulated annealing, with much shorter computation time [125].

Color Permutation

The objective of the color permutation problem is to maximize the layer separation between interfering nets. An effective algorithm for this problem is presented in [126], based on the max-cut partitioning algorithm described above.

The algorithm is based on the following fact: each of the $k-1$ cuts in the maximal color tree generated by the maxcut partitioning algorithm defines a "global partition" in the color permutation problem. For example, suppose the first cut in the maximal color tree is known, which separates $V(G)$ into X_1 and X_2. The above statement implies that in an optimal solution to the color permutation problem, all vertices belonging to X_1 must lie on the same side of all vertices belonging to X_2.

If we know a partition in the maximal k-color ordering, for example, the cut separating the leftmost $n/2$ vertices from the rightmost $n/2$ vertices, then the number of possible permutations is reduced from $n!$ to $(n/2! \times n/2!)$. Motivated by the above fact, the color partitioning algorithm is built on the same binary tree partitioning structure as used for the max-cut partitioning problem. The intuition behind the algorithm is that the set of global partitions induced by a series of maximum cuts significantly reduces the permutation configuration size at each stage of the binary decomposition. At each level of the color tree, the best permutation is determined for the set of subgraphs in the level. Then, the set of subgraphs rearranged with the best cost at level i is further decomposed into smaller subgraphs whose optimal ordering is considered in level $i+1$ of the top-down color tree. Figure 6.13 illustrates a color tree with three stages of partitioning and the number of possible permutations at the second stage.

Based on the above approach, an approximation algorithm for the permutation problem can be devised which runs in $O(|E|k^2 + k^3)$ time yielding the following lower bound on the *total weighted net separation*, S:

$$S \geq \left(\frac{k^2 - 1}{3k}\right) w(E) \tag{6.9}$$

where $w(E) = \sum_{e \in E} w(e)$. For example, when $k = 4$, $S \geq 5w(E)/4$. A proof of the bound and a detailed description of the algorithm can be found in [125].

6.4   SUMMARY

The problem of layer assignment in multilayer MCMs was considered. In the first part of the chapter, a model for the multilayer substrate in a ceramic MCM was developed, and a new constrained layer assignment problem was formulated on this model. The second part of the chapter considered a different approach, in which a net interference graph (NIG) is generated from a netlist or global routing, and the NIG is partitioned into a given number of subsets, with the

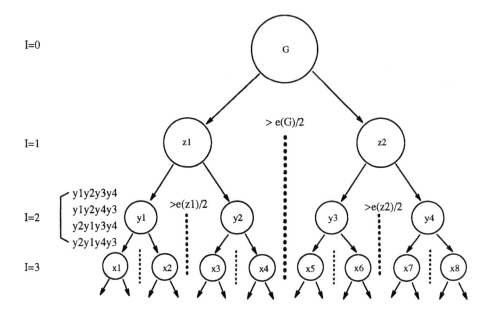

Figure 6.13 A maxcut color-tree with color permutation

objective of maximizing the planarity of nets in each subset. This approach is applicable to unconstrained as well as constrained layer assignments, and can be used in a gridless routing environment also. Crosstalk minimization was also considered, and new approaches for constructing the NIG were presented. New algorithms for the max-cut partitioning and layer permutation problems were briefly described.

7

CONCLUSIONS

7.1 SUMMARY

The role of packaging has grown dramatically in importance in the past few years, and high-performance packages such as multichip modules have become increasingly popular. Careful design of these packages can significantly increase the benefits they provide.

A considerable body of literature on physical design problems in VLSI exists. The first objective of this book was to provide the reader with a clear understanding of what an MCM is, and how it differs physically from a VLSI chip. Chapter 1 gave an overview of the various technologies, features and design problems associated with MCMs, and provided the background for the material discussed in the following chapters.

The physical design of a VLSI chip involves a large number of active and passive devices and interconnections, often with variable sizes, shapes and locations which can be adjusted for optimum performance. In contrast, MCM physical design is focussed primarily on *interconnections*. Thus, interconnections form a natural starting point for the study of MCM physical design, and are discussed at length in Chapter 2. Unlike on-chip wires, which can be modeled fairly well as lumped capacitors, interchip wires have to be modeled as lossy transmission lines, which are notoriously difficult to analyze. The computational cost of accurate timing analysis of all the nets in an MCM can easily dominate the cost of the entire physical design process. New approaches to interconnect analysis are presented in Chapter 2, which can compute approximate time-domain responses of MCM interconnects. The rapid estimation approaches can provide excellent approximations of the time-domain response, while running as much as several hundred times faster than a conventional simulation program.

Chapter 2 also describes delay models for MCM interconnects. A new second-order delay model is developed for multiterminal interconnects, which is able to capture the effect of line inductance on net delays. Existing delay models are limited to RC models, which can be very optimistic when transmission line effects come into play.

The problems of system partitioning and chip placement are considered in Chapter 3. The partitioning and placement problems for VLSI have been studied intensively for several years, and several effective algorithms have been developed. However, none of these algorithms can be applied directly to MCMs, since they usually ignore timing constraints, or use delay models which are not appropriate for MCM interconnects. On-chip wires are short, and their resistance is usually negligible compared to the output impedance of the driver. However, MCM wires are much longer, and the output impedances of off-chip drivers are smaller than those of on-chip drivers. Thus, wire resistance cannot be ignored in performance-driven design of MCMs. This makes the MCM partitioning and placement problems much more complicated, since net delays depend on the actual net *topology*, in addition to the wire length. Chapter 3 describes timing-driven partitioning algorithms which are able to handle arbitrary delay models, and a *resistance-driven* placement algorithm, which takes wire resistance into account to find a placement minimizing interchip net delays. This algorithm is demonstrated to be capable of generating placements with significantly smaller net delays than conventional approaches, which ignore wire resistance.

The physical design step which follows placement is routing. MCM routing may involve many layers of wiring, unlike VLSI, where two or three layers usually suffice. In addition, MCM routing has unique features, such as segmented vias, and complex constraints such as transmission line effects, which make it necessary to use new routing algorithms developed specifically for MCMs. Chapter 4 describes some recent algorithms capable of handling the complexity of multilayer MCM routing.

Chapter 5 describes another popular approach to MCM routing, in which the multilayer problem is decomposed into two stages: two-dimensional routing and layer assignment. For the two-dimensional tree-generation problem, the conventional minimum Steiner tree heuristics used in VLSI are not adequate, since they ignore both wire resistance and inductance. Recent work on performance-oriented tree construction algorithms is presented, in which trees are constructed for minimizing first- or second-order delays.

Chapter 6 deals with the second stage of multilayer routing, layer assignment. Two approaches to the problem are considered. The first approach is based on a new model for the MCM-C multilayer routing environment, which allows such features as segmented vias and terminals on the top as well as the bottom layers. The second approach does not require a routing grid, and is hence more suitable for high-density MCMs, such as MCM-Ds. It is based on a graph partitioning problem on an "interference graph" generated from a netlist or a two-dimensional global routing. Techniques for constructing the interference graph and for minimizing crosstalk between nets are described.

7.2 OPEN PROBLEMS FOR FUTURE WORK

Whereas physical design algorithms for VLSI design have been studied intensively for over two decades, papers on MCM physical design have begun to appear in journals and conferences only over the past two or three years. Thus, a number of interesting problems remain open for future research in this area. Some of these are described below:

Rapid interconnect simulation: A number of rapid interconnect simulation approaches have been proposed recently, such as asymptotic waveform evaluation [34], reciprocal expansion, complex frequency hopping [127], and approaches based on recursive convolution [27]. A detailed comparative study of these different approaches, with theoretical results and experimental data, would be a very valuable contribution.

Interconnect delay modeling: The first-order delay models commonly used in performance-driven design are not adequate for MCMs. Even the second-order delay model presented in Chapter 2 accounts for a single mode of oscillation in the interconnect response, namely, the oscillations caused by a mismatched driver impedance. In practice, however, the settling delay may be increased by oscillatory modes introduced by stubs. A delay model which is able to predict the effect of stubs without having to compute a higher-order approximation of the interconnect would be very useful for physical design.

Performance-driven placement: An extension of the resistance-driven placement algorithm to an RLC-driven placement algorithm would further reduce interchip signal delays, by taking the interconnect inductance into account. Thermal constraints also have to be taken into account more carefully in MCM placement.

Tree construction algorithms: Research in Steiner tree algorithms is moving away from the traditional minimum wire-length objective, and is becoming more performance oriented. More sophisticated algorithms for minimizing second-order delays have to be developed. One of the algorithms presented in Chapter 5 attempts to reduce the total number and lengths of stubs. This approach could be made more effective in conjunction with a good model for the effect of stubs on settling delays.

Crosstalk avoidance has not received much attention during the tree construction stage – trees are typically constructed one at a time, independently of other trees. More effective tree construction algorithms can be developed, which take into account the locations of other nets to construct a tree with minimum RLC delay and crosstalk.

Layer assignment: The layer assignment strategy based on max-cut graph partitioning has to be developed further. In particular, better measures for nonplanarity have to be developed. A "hybrid" multilayer routing strategy, which combines the advantages of the constrained and unconstrained layer assignment strategies, is also worth investigating.

A systematic investigation of the relative performance of various multilayer routing approaches – 3-D maze routing, layer-by layer approaches such as SLICE and V4R, constrained and unconstrained multilayer routing – would give a clear picture of which approach is most likely to produce better results. This would allow researchers to focus their attention on a few selected strategies and maximize their effectiveness. Theoretical and experimental comparisons of unidirectional wiring on $x - y$ plane pairs and bidirectional wiring layers can also provide industry with valuable information on technology options.

Optical interconnects: As system speeds approach the gigahertz range, optical interconnects will become essential. Since the nature of the physical constraints in an optical interconnect system will be quite different from the conventional electrical interconnects, optoelectronic MCMs will introduce a new set of physical design problems. Some work on layout design for optoelectronic MCMs has already been reported [128].

Packaging technology selection: The multitude of packaging options available today can make it very difficult for system designers to choose the best package for their design. The packaging parameters interact with each other in complex ways, and their impact on the system size, performance, cost, reliability and manufacturability is very difficult to predict. Choosing an appropriate package requires expertise in a wide range of fields, including material science, electrical engineering, computer science and mechanical engineering. Integrating this expertise into a CAD tool for package selection can make it possible for a small team of designers to quickly select a package which meets the system-level requirements. Such tools are under development now, and will continue to be refined as cost and performance models improve.

REFERENCES

[1] A. Mehta et al. SuperSPARC multi chip module. In *Proc. IEEE Multi-Chip Module Conference*, pages 19–28, March 1993.

[2] A. J. Blodgett and D. R. Barbour. Thermal conduction module: A high-performance ceramic package. *IBM Journal of Research and Development*, 26:30–36, 1982.

[3] R. R. Tummala and E. J. Rymaszewski. *Microelectronics Packaging Handbook*. Van Nostrand Reinhold, 1989.

[4] J. U. Knickerbocker et al. IBM System/390 air cooled alumina thermal conduction module. *IBM Journal of Research and Development*, 35(3):330–341, May 1991.

[5] D. H. Carey. Trends in low-cost, high-performance substrate technology. *IEEE Micro*, pages 19–27, April 1993.

[6] W. Daum, W. E. Burdick Jr., and R. A. Fillion. Overlay high-density interconnect: A chips-first multichip module technology. *IEEE Computer*, pages 23–29, April 1993.

[7] T. Costlow. Two join MCM market. *Electronic Engineering Times*, page 58, June 7 1993.

[8] R. K. Scannell and J. K. Hagge. Development of a multichip module DSP. *IEEE Computer*, pages 13–21, April 1993.

[9] P. D. Franzon and R. J. Evans. A multichip module design process for notebook computers. *IEEE Computer*, pages 41–49, April 1993.

[10] Mentor Graphics. Grumman leverages MCM design for satellite electronics. *Mentor Graphics Vision*, pages 12–13, Autumn 1992.

[11] R. W. Broderson and W. B. Baringer. MCMs for portable applications. In *Proc. IEEE Multi-Chip Module Conference*, pages 1–5, Santa Cruz, CA, March 1993.

[12] D. P. LaPotin. Early assessment of design, packaging and technology tradeoffs. *Intl. Journal of High Speed Electronics*, 2(4):209–233, 1991.

[13] P. A. Sandborn. A software tool for technology tradeoff evaluation in multichip packaging. In *IEEE/CHMT IEMT Symposium*, pages 337–340, 1991.

[14] P. H. Dehkordi and D. W. Bouldin. Design for packagability: The impact of bonding technology on the size and layout of VLSI dies. In *Proc. IEEE Multi-Chip Module Conference*, pages 153–159, Santa Cruz, CA, March 1993.

[15] T. Gabara, W. Fischer, S. Knauer, R. Frye, K. Tai, and M. Lau. An I/O CMOS buffer set for silicon multi chip modules. In *Proc. IEEE Multi-Chip Module Conference*, pages 147–152, Santa Cruz, CA, March 1993.

[16] M. Sriram. *Physical Design for Multichip Modules*. PhD dissertation, University of Illinois at Urbana-Champaign, Department of Electrical and Computer Engineering, October 1993.

[17] D. Zhou, F. Tsui, J. S. Cong, and D. S. Gao. A distributed-RLC model for MCM layout. In *Proc. IEEE Multi-Chip Module Conference*, pages 191–197, Santa Cruz, CA, March 1993.

[18] M. Sriram and S. M. Kang. Performance driven MCM routing using a second order RLC tree delay model. In *Proc. IEEE Intl. Conf. on Wafer Scale Integration*, pages 262–267, San Francisco, January 1993.

[19] W. W.-M. Dai. Multichip routing and placement. *IEEE Spectrum*, pages 61–64, November 1992.

[20] C. E. Leiserson and F. M. Maley. Algorithms for routing and testing routability of planar VLSI layouts. In *Proc. 17th Annual ACM Symposium on Theory of Computing*, pages 69–78, 1985.

[21] K.-Y. Khoo and J. Cong. A fast multilayer general area router for MCM designs. *IEEE Transactions on Circuits and Systems*, pages 841–851, November 1992.

[22] C.-H. Chen, M. H. Heydari, I. G. Tollis, and C. Xia. Improved layer assignment for packaging multichip modules. Technical Report UTDCS-14-92, UT Dallas, September 1992.

[23] M. Sriram and S. M. Kang. Detailed layer assignment for MCM routing. In *Dig. Tech. Papers, Intl. Conf. on Computer-Aided Design*, pages 386–389, November 1992.

[24] J. M. Ho, M. Sarrafzadeh, G. Vijayan, and C. K. Wong. Layer assignment for multi-chip modules. *IEEE Transactions on Computer-Aided Design*, 9(12):1272–1277, 1990.

[25] C. L. Ratzlaff, N. Gopal, and L. T. Pillage. RICE: Rapid interconnect circuit evaluator. In *Proc. ACM/IEEE Design Automation Conference*, pages 555–560, June 1991.

[26] M. Sriram and S. M. Kang. Efficient approximation of the time domain response of lossy coupled transmission line trees. In *Proc. ACM/IEEE Design Automation Conference*, pages 691–696, June 1993.

[27] S. Lin and E. S. Kuh. Transient simulation of lossy interconnect. In *Proc. ACM/IEEE Design Automation Conference*, pages 81–86, June 1992.

[28] M. Sriram and S. M. Kang. iPROMIS: An interactive performance driven multilayer MCM router. In *Proc. IEEE Multi-Chip Module Conference*, pages 170–173, March 1993.

[29] Q. J. Zhang and M. Nakhla. Yield analysis and optimization of VLSI interconnects in multichip modules. In *Proc. IEEE Multi-Chip Module Conference*, pages 160–163, March 1993.

[30] S. Burman and N. A. Sherwani. Programmable multichip modules. *IEEE Micro*, pages 28–35, April 1993.

[31] K. W. Kelley, G. T. Valliath, and J. W. Stafford. High-speed chip-to-chip optical interconnect. *IEEE Photonics Technology Letters*, 4(10):1157–1159, October 1992.

[32] H. B. Bakoglu. *Circuits, Interconnections and Packaging for VLSI*. Addison-Wesley, 1990.

[33] J. S. Roychowdhury, A. R. Newton, and D. O. Pederson. Simulating lossy interconnect with high frequency nonidealities in linear time. In *Proc. ACM/IEEE Design Automation Conference*, pages 75–80, 1992.

[34] L. Pillage and R. Rohrer. Asymptotic waveform evaluation for timing analysis. *IEEE Transactions on Computer-Aided Design*, 9(4):352–366, April 1990.

[35] V. Raghavan, R. A. Rohrer, L. T. Pillage, J. Y. Lee, J. E. Bracken, and M. M. Alaybeyi. Awe-inspired. In *Custom Integrated Circuits Conference*, pages 18.1.1–18.1.8, May 1993.

[36] G. A. Baker Jr. *Essentials of Padé Approximants*. Academic Press, 1975.

[37] T. Dhaene and D. De Zutter. Selection of lumped element models for coupled lossy transmission lines. *IEEE Transactions on Computer-Aided Design*, 11(7):805–815, July 1992.

[38] T. K. Tang and M. Nakhla. Analysis of high speed VLSI interconnects using the asymptotic waveform evaluation technique. In *Dig. Tech. Papers, Intl. Conf. on Computer-Aided Design*, pages 542–545, 1990.

[39] E. Bracken, V. Raghavan, and R. A. Rohrer. Simulating distributed elements with asymptotic waveform evaluation. In *Proc. IEEE International Microwave Symposium*, June 1992.

[40] N. Gopal and L. Pillage. Constrained approximation of dominant time constants in RC circuit delay models. Technical Report UT-CERC-TR-LTP91-01, University of Texas at Austin, November 1991.

[41] D. Xie and M. Nakhla. Delay and crosstalk simulation of high-speed VLSI interconnects with nonlinear terminations. In *Dig. Tech. Papers, Intl. Conf. on Computer-Aided Design*, pages 66–69, November 1991.

[42] E. Chiprout and M. Nakhla. Addressing high-speed interconnect issues in Asymptotic Waveform Evaluation. In *Proc. ACM/IEEE Design Automation Conference*, pages 732–736, June 1993.

[43] E. Chiprout and M. Nakhla. Transient waveform estimation of high-speed MCM networks using complex frequency hopping. In *Proc. IEEE Multi-Chip Module Conference*, pages 134–139, March 1993.

[44] S. P. McCormick and J. Allen. Waveform moment methods for improved interconnection analysis. In *Proc. ACM/IEEE Design Automation Conference*, pages 406–412, June 1990.

[45] J. S. Roychowdhury and D. O. Pederson. Efficient transient simulation of lossy interconnect. In *Proc. ACM/IEEE Design Automation Conference*, pages 740–745, June 1991.

[46] D. Zhou, S. Su, F. Tsui, D. S. Gao, and J. Cong. Analysis of trees of transmission lines. Technical Report CSD-920010, UCLA, 1992.

[47] C. S. Yen, Z. Fazarinc, and R. L. Wheeler. Time-domain skin effect model for transient analsis of lossy transmission lines. *Proceedings of the IEEE*, 70(7):750–757, July 1982.

[48] D. Zhou, F. P. Preparata, and S. M. Kang. Interconnection delay in very high-speed VLSI. *IEEE Transactions on Circuits and Systems*, 38(7):779–790, July 1991.

[49] W. C. Elmore. The transient response of damped linear networks with particular regard to wideband amplifiers. *Journal of Applied Physics*, 19(1):55–63, January 1948.

[50] J. Rubinstein, P. Penfield, and N. A. Horowitz. Signal delay in RC tree networks. *IEEE Transactions on Computer-Aided Design*, 2(3):202–211, 1983.

[51] L. R. Ford and D. R. Fulkerson. *Flows in Networks*. Princeton University Press, Princeton, New Jersey, 1962.

[52] B. W. Kernighan and S. Lin. An efficient heuristic procedure for partitioning graphs. *The Bell System Technical Journal*, 49(2):291–307, February 1970.

[53] C. M. Fiduccia and R. M. Mattheyses. A linear time heuristic for improving network partitions. In *Proc. ACM/IEEE Design Automation Conference*, pages 175–181, 1982.

[54] Y.-C. Wei and C. K. Cheng. Ratio cut partitioning for hierarchical designs. *IEEE Transactions on Computer-Aided Design*, 10(7):911–921, July 1991.

[55] J. Cong, L. Hagen, and A. Kahng. Net partitions yield better module partitions. In *Proc. ACM/IEEE Design Automation Conference*, pages 47–52, June 1992.

[56] N. R. Quinn and M. A. Breuer. A force-directed component placement procedure for printed circuit boards. *IEEE Transactions on Circuits and Systems*, CAS-26:377–388, 1979.

[57] S. Kirkpatrick, C. D. Gelatt, and M. P. Vecchi. Optimization by simulated annealing. *Science*, 220:671–680, 1983.

[58] C. Sechen. *VLSI Placement and Global Routing Using Simulated Annealing*. Kluwer Academic Publishers, Boston, MA, 1988.

[59] R. M. Kling and P. Banerjee. Empirical and theoretical studies of the simulated evolution method applied to standard cell placement. *IEEE Transactions on Computer-Aided Design*, 10(10):1303–1315, 1991.

[60] M. A. Breuer. Min-cut placement. *Journal on Design Automation and Fault-Tolerant Computing*, 1:343–382, October 1977.

[61] M. Shih, E. S. Kuh, and R.-S. Tsay. Performance-driven system partitioning on multi-chip modules. In *Proc. ACM/IEEE Design Automation Conference*, pages 53–56, June 1992.

[62] M. Shih and E. S. Kuh. Quadratic Boolean programming for performance-driven system partitioning. In *Proc. ACM/IEEE Design Automation Conference*, pages 761–765, June 1993.

[63] M. Shih, E. S. Kuh, and R.-S. Tsay. Performance-driven system partitioning on multi-chip modules. Technical Report RC 17315 (#76556), IBM, October 1991.

[64] P. S. Hauge, R. Nair, and E. J. Yoffa. Circuit placement for predictable performance. In *Dig. Tech. Papers, Intl. Conf. on Computer-Aided Design*, pages 88–91, November 1987.

[65] H. Youssef and E. Shragowitz. Timing constraints for correct performance. In *Dig. Tech. Papers, Intl. Conf. on Computer-Aided Design*, pages 24–27, November 1990.

[66] J. Frankle. Iterative and adaptive slack allocation for performance-driven layout and FPGA routing. In *Proc. ACM/IEEE Design Automation Conference*, pages 536–542, June 1992.

[67] M. A. B. Jackson and E. S. Kuh. Performance-driven placement of cell-based ICs. In *Proc. ACM/IEEE Design Automation Conference*, pages 370–375, 1989.

[68] T. Gao, P. M. Vaidya, and C. L. Liu. A new performance-driven placement algorithm. In *Dig. Tech. Papers, Intl. Conf. on Computer-Aided Design*, pages 44–47, 1991.

[69] M. Terai, K. Takahashi, and K. Sato. A new min-cut placement algorithm for timing assurance layout design meeting net length constraints. In *Proc. ACM/IEEE Design Automation Conference*, pages 96–102, 1990.

[70] A. Srinivasan, K. Chaudhary, and E. S. Kuh. RITUAL: A performance-driven placement algorithm. *IEEE Transactions on Circuits and Systems*, 39(11):825–840, November 1992.

[71] P. R. Suaris and G. Kedem. A quadrisection-based combined place and route scheme for standard cells. *IEEE Transactions on Computer-Aided Design*, 8(3):234–244, 1989.

[72] M. Burstein. A non 'placement-routing' approach to automation of VLSI layout design. In *Proc. Intl. Symposium on Circuits and Systems*, pages 756–759, 1982.

[73] W.-M. Dai and E. S. Kuh. Simultaneous floorplanning and global routing for hierarchical building block layout. *IEEE Transactions on Computer-Aided Design*, CAD-6:828–837, 1987.

[74] J. Cong, K.-S. Leung, and D. Zhou. Performance-driven interconnect design based on distributed RC delay model. Technical Report CSD-920043, UCLA, October 1992.

[75] R. B. Hitchcock, G. L. Smith, and D. D. Cheng. Timing analysis of computer hardware. *IBM Journal of Research and Development*, 26(1):100–105, January 1982.

[76] W. E. Donath et al. Timing driven placement using complete path delays. In *Proc. ACM/IEEE Design Automation Conference*, pages 84–89, 1990.

[77] P. R. Suaris and G. Kedem. An algorithm for quadrisection and its application to standard cell placement. *IEEE Transactions on Circuits and Systems*, 35(3):294–303, 1988.

[78] A. E. Dunlop and B. W. Kernighan. A procedure for layout of standard cell VLSI circuits. *IEEE Transactions on Computer-Aided Design*, CAD-4:92–98, 1985.

[79] C. Sechen and D. Chen. An improved objective function for mincut circuit partitioning. In *Dig. Tech. Papers, Intl. Conf. on Computer-Aided Design*, pages 502–505, 1988.

[80] A. Hanafusa, Y. Yamashita, and M. Yasuda. Three-dimensional routing for multilayer ceramic printed circuit boards. In *Dig. Tech. Papers, Intl. Conf. on Computer-Aided Design*, pages 386–389, November 1990.

[81] K.-Y. Khoo and J. Cong. An efficient multilayer MCM router based on four-via routing. In *Proc. ACM/IEEE Design Automation Conference*, pages 590–595, June 1993.

[82] W. W.-M. Dai, R. Kong, J. Jue, and M. Sato. Rubber band routing and dynamic data representation. In *Dig. Tech. Papers, Intl. Conf. on Computer-Aided Design*, November 1990.

[83] F. P. Preparata and M. I. Shamos. *Computational Geometry*. Springer-Verlag, New York, NY, 1985.

[84] C. L. Liu. *Elements of Discrete Mathematics*. McGraw-Hill, New York, NY, 1977.

[85] W. W.-M. Dai. Performance driven layout of thin-film substrates for multichip modules. In *Multichip Module Workshop Proceedings*, pages 114–121, March 1991.

[86] W. W.-M. Dai, R. Kong, and M. Sato. Routability of a rubber band sketch. In *Proc. ACM/IEEE Design Automation Conference*, pages 45–48, June 1991.

[87] W. W.-M. Dai, T. Dayan, and D. Staepelaere. Topological routing in SURF: Generating a rubber band sketch. In *Proc. ACM/IEEE Design Automation Conference*, pages 39–44, June 1991.

[88] L. P. Chew. Constrained Delauney triangulation. *Algorithmica*, 4:97–108, 1989.

[89] X.-M. Xiong and E. S. Kuh. The constrained via minimization problem for PCB and VLSI design. In *Proc. ACM/IEEE Design Automation Conference*, pages 573–578, 1988.

[90] C.-P. Hsu. Minimum-via topological routing. *IEEE Transactions on Computer-Aided Design*, CAD-2(4):235–246, October 1983.

[91] M. Sarrafzadeh and D. T. Lee. A new approach to topological via minimization. *IEEE Transactions on Computer-Aided Design*, 8(8):890–900, August 1989.

[92] A. Lim, S.-W. Cheng, and C.-T. Wu. Performance oriented rectilinear Steiner trees. In *Proc. ACM/IEEE Design Automation Conference*, pages 171–176, June 1993.

[93] X. Hong, T. Xue, E. S. Kuh, C.-K. Cheng, and J. Huang. Performance driven Steiner tree algorithms for global routing. In *Proc. ACM/IEEE Design Automation Conference*, pages 177–181, June 1993.

[94] K. D. Boese, A. B. Kahng, and G. Robins. High-performance routing trees with identified critical sinks. In *Proc. ACM/IEEE Design Automation Conference*, pages 182–187, June 1993.

[95] K. D. Boese, A. B. Kahng, B. A. McCoy, and G. Robins. Fidelity and near-optimality of Elmore-based routing constructions. Technical Report CS-93-14, Dept. of CS, Univ. of Virginia, Charlottesville, 1993.

[96] S. Prasitjutrakul and W. Kubitz. A timing-driven global router for custom chip design. In *Dig. Tech. Papers, Intl. Conf. on Computer-Aided Design*, pages 48–51, 1990.

[97] N. J. Nilsson. *Problem-Solving Methods in Artificial Intelligence.* McGraw-Hill, 1971.

[98] K. D. Boese, A. B. Kahng, and G. Robins. High-performance routing trees with identified critical sinks. In *Proc. ACM/IEEE Design Automation Conference*, pages 182–187, June 1993.

[99] J. Cong, A. Kahng, G. Robins, M. Sarrafzadeh, and C. K. Wong. Provably good performance-driven global routing. *IEEE Transactions on Computer-Aided Design*, 11(6):739–752, June 1992.

[100] A. Lim, S.-W. Cheng, and C.-T. Wu. Performance oriented rectilinear Steiner trees. In *Proc. ACM/IEEE Design Automation Conference*, pages 171–176, June 1993.

[101] X. Hong, T. Xue, E. S. Kuh, C.-K. Cheng, and J. Huang. Performance-driven Steiner tree algorithms for global routing. In *Proc. ACM/IEEE Design Automation Conference*, pages 177–181, June 1993.

[102] S. K. Rao, P Sadayappan, F. K. Hwang, and P. W. Shor. The rectilinear Steiner arborescence problem. *Algorithmica*, 7:277–288, 1992.

[103] M. Sarrafzadeh and C. K. Wong. Hierarchical Steiner tree construction in uniform orientations. *IEEE Transactions on Computer-Aided Design*, pages 1095–1103, September 1992.

[104] K. F. Liao, M. Sarrafzadeh, and C. K. Wong. Single-layer global routing. Technical Report RC 16900 (#74940), IBM, May 1991.

[105] D. F. Wann and M. A. Franklin. Asynchronous and clocked control structures for VLSI based interconnection networks. *IEEE Transactions on Computers*, c-32(3):284–293, March 1983.

[106] M. A. B. Jackson, A. Srinivasan, and E. S. Kuh. Clock routing for high performance ICs. In *Proc. ACM/IEEE Design Automation Conference*, pages 573–579, June 1990.

[107] A. Kahng, J. Cong, and G. Robins. High-performance clock routing based on recursive geometric matching. In *Proc. ACM/IEEE Design Automation Conference*, pages 322–327, June 1991.

[108] K. J. Supowit and E. M. Reingold. Divide and conquer heuristics for minimum weighted Euclidean matching. *SIAM J. Computing*, 12(1):118–143, 1983.

[109] R.-S. Tsay. Exact zero skew. In *Dig. Tech. Papers, Intl. Conf. on Computer-Aided Design*, pages 336–339, November 1991.

[110] T.-H. Chao, Y.-C. Hsu, and J.-M. Ho. Zero skew clock net routing. In *Proc. ACM/IEEE Design Automation Conference*, pages 518–523, June 1992.

[111] S. Pullela, N. Menezes, and L. T. Pillage. Reliable non-zero skew clock trees using wire width optimization. In *Proc. ACM/IEEE Design Automation Conference*, pages 165–170, June 1993.

[112] L. P. P. P. van Ginneken. Buffer placement in distributed RC tree networks for minimal Elmore delay. In *Proc. Intl. Symposium on Circuits and Systems*, pages 865–868, May 1990.

[113] J. D. Cho and M. Sarrafzadeh. A buffer redistribution algorithm for high-speed clock routing. In *Proc. ACM/IEEE Design Automation Conference*, pages 537–542, June 1993.

[114] M. Marek-Sadowska. An unconstrained topological via minimization problem for two-layer routing. *IEEE Transactions on Computer-Aided Design*, CAD-3(3):184–190, July 1984.

[115] K. C. Chang and H. C. Du. Layer assignment problem for three-layer routing. *IEEE Transactions on Computers*, 37(5):625–632, May 1988.

[116] F. Hadlock. Finding a maximum cut of a planar graph in polynomial time. *SIAM Journal on Computers*, 4:221–225, September 1975.

[117] L. C. Abel. On the automated layout of multilayer planar wiring and a related graph-coloring problem. Technical Report R-546, Coordinated Science Laboratory, University of Illinois at Urbana-Champaign, 1972.

[118] H. C. So. Some theoretical results on the routing of multilayer printed wiring boards. In *Proc. Intl. Symposium on Circuits and Systems*, pages 296–303, 1974.

[119] R. Raghavan and S. Sahni. Some complexity results on the single-row approach to wiring. In *Proc. Intl. Symposium on Circuits and Systems*, pages 768–771, 1982.

[120] T. Tarng, M. Marek-Sadowska, and E. S. Kuh. An efficient single-row routing algorithm. *IEEE Transactions on Computer-Aided Design*, CAD-3(3):178–183, July 1984.

[121] A. Aggarwal, M. Klawe, and P. Shor. Multilayer grid embedding for VLSI. *Algorithmica*, 6(1):129–151, 1991.

[122] B. S. Ting and E. S. Kuh. An approach to the routing of multilayer printed circuit boards. In *Proc. Intl. Symposium on Circuits and Systems*, pages 902–911, 1978.

[123] M. Sriram and S. M. Kang. A new layer assignment approach for mcms. Technical Report UIUC-BI-VLSI-92-01, University of Illinois at Urbana-Champaign, March 1992.

[124] J. D. Cho, Northwestern University. Private communication.

[125] J. D. Cho, S. Raje, and M. Sarrafzadeh. Approximation for the maximum cut, k-coloring and maximum linear arrangement problems. Manuscript, EECS Dept., Northwestern University, 1993.

[126] S. Raje, J. D. Cho, M. Sarrafzadeh, M. Sriram, and S. M. Kang. Crosstalk-minimum layer assignment. In *IEEE Custom Integrated Circuits Conference*, San Diego, May 1993.

[127] E. Chiprout and M. Nakhla. Addressing high-speed interconnect issues in asymptotic waveform evaluation. In *Proc. ACM/IEEE Design Automation Conference*, pages 732–736, June 1993.

[128] J. Fan, D. Zaleta, C. K. Cheng, and S. H. Lee. Physical layout algorithms for computer generated holograms in optoelectronic MCM systems design. In *Proc. IEEE Multi-Chip Module Conference*, pages 170–173, March 1993.

INDEX

Printed in the USA
CPSIA information can be obtained
at www.ICGtesting.com
LVHW050058111023
760655LV00007B/221